Monographs on Endocrinology

Volume 1

Edited by

A. Labhart, Zürich · T. Mann, Cambridge

L. T. Samuels, Salt Lake City · J. Zander, Heidelberg

Susumu Ohno

Sex Chromosomes and Sex-linked Genes

With 33 Figures

Springer-Verlag Berlin · Heidelberg · New York 1967

Dr. *Susumu Ohno*
Department of Biology
City of Hope Medical Center
Duarte/Calif./USA

This book is dedicated to my teacher of many years, Professor Riojun Kinosita in commemoration of his seventy-third birthday

Preface

There is an Arabian proverb which says that the magnificient Arabian steed and the ungainly camel are actually created of the same material. While one god was able to create the horse, the same material in the hands of a committee of gods wound up as the camel.

On the premise that each field of natural science has become too complex to be comprehended by a single man, it is more fashionable today to organize a committee of specialized scientists to write one book. While a book written by a committee tends to present an objective appraisal of current knowledges, it suffers from disunity of thoughts. It is my sincere desire that this book will manifest more merits than shortcomings in having been written by one author.

SUSUMU OHNO

Acknowledgements

First of all, I am grateful to my many colleagues with whom I have had the pleasure of engaging in various cooperative ventures. The results of these ventures form the foundation of this book.

I am also grateful to Mrs. LENORE ANDERSEN and Mrs. MAXINE WALKER for the help in preparing this book.

My gratitude is also to my wife and children who have endured the negligence of six months during the preparation of this book.

The frank criticism of this manuscript offered by Drs. MELVIN COHN, ALFRED G. KNUDSON and ERNEST BEUTLER are greatly appreciated.

This work was supported in part by grant CA-05138-07 from the National Cancer Institute, U.S. Public Health Service, and in part by a research fund established in honor of General James H. Doolittle. Contribution No. 4-67, Department of Biology, City of Hope Medical Center.

Contents

Part III

On So-called Sex Determining Factors and the Act of Sex Determination

Introduction

Natural selection operates among individual organisms which differ in their genetic constitution. The degree of hereditary variability within a species is greatly enhanced by cross-fertilization. Indeed, the mechanism of sexual reproduction occurred very early in evolution, for it is seen today even in bacteria. In *Escherichia coli,* fertilization occurs by passage of the single chromosome from the male into the female bacterium (LEDERBERG, 1959).

In multicellular organisms, the separation of germ from soma, and the production of haploid gametes became mandatory. The gametes were of two types. One, extremely mobile, was designed to seek out and penetrate the other, which loaded with nutrients, received the mobile gamete and intiated the development of a new individual. The foundation for true bisexuality was thus laid.

In the primitive state of bisexuality, whether an individual is to be a sperm-producing male or an egg-producing female appears to be decided rather haphazardly. In the worm, *Bonelia viridis,* the minute males are parasites in the female. Larvae that become attached to the proboscis of an adult female become males, while unattached larvae sink to the bottom and become females (BALTZER, 1935).

The more sophisticated state of bisexuality was initiated by setting aside a particular pair of chromosomes for specialization and making either the male or the female a heterogametic sex. Sex chromosomes as we know them were thus born.

Among surviving members of vertebrates, the truly differentiated X- and Y-chromosomes occur only in mammals, while female hetero-gamety with well-differentiated Z- and W-chromosomes operates in avian and ophidian species. At the primitive extreme, it appears that certain fish species of today are still without a chromosomal sex-determining mechanism. The entire spectrum of the chromosomal sex-determining mechanism with sex chromosomes in varying degrees of differentiation is indeed exhibited by surviving vertebrates. This monograph has been written so that the step-by-step reconstruction of the phylogenetic past of the sex chromosomes in relation to the evolution of vertebrate genomes might broaden our understanding of the nature of sex-determining factors, sex-linked genes, and the dosage compensation mechanism.

Part I deals with the evolution of vertebrate sex chromosomes within the context of polyphyletic evolution of vertebrate genomes. It will be shown that the X and the Y, or the Z and the W, were originally an homologous pair of ordinary chromosomes. Subsequent differentiation of this pair to the sex pair was accomplished exclusively at the expense of one member of the pair which was elected to accumulate the factors governing the development of the heterogametic sex, the Y-chromosome of the male heterogamety, and the W of the female heterogamety. Although the Y or the W has eliminated all the Mendelian genes which were originally on it by progressive genetic deterioration and finally emerged as the very specialized sex determiner of minute size, no substantial change occurred to the other member of the pair which was elected to accumulate the determining factors of the homogametic sex. Instead, it was conserved in its entirety as the X or the Z. The so-called sex-linked genes are nothing more than the original Mendelian genes which were there when the X or the Z was an ordinary chromosome.

In regard to the evolution of vertebrate genomes, evidence will be presented indicating that a series of polyploidization of the ancestral genome occurred in ancient times when vertebrates were still aquatic. As a result, crossopterygian fishes which evolved to terrestrial vertebrates represented diverse genome lineages in varying degrees of polyploidization. The subsequent emergence of the chromosomal sex-determining mechanism served to stabilize each genome lineage, since further polyploidization would have led to intersexuality. Placental mammals as a group belong to one genome lineage, and avian species to another. The total DNA content is the same for diverse species of placental mammals. Evidently, speciation within the infraclass, *Eutheria,* has been accomplished mainly by allelic mutations coupled with constant reshuffling of the autosomal linkage groups. While the autosomal linkage groups were broken and reunited many times, the original X of a common ancestor was apparently preserved in its entirety. So far as the X-linked genes of placental mammals are concerned, it appears that whatever is X-linked in one species is also X-linked in others.

The reptilian suborder, *Squamata,* and the class, *Aves,* belong to one genome lineage with a total DNA content of about 50% of that of placental mammals. The female heterogamety of the ZZ/ZW-type operates in this genome lineage. The original Z of this lineage was also preserved in its entirety by diverse ophidian as well as avian species.

Part II deals with the dosage effect of sex-linked genes and the need for an effective dosage compensation mechanism. In fishes, amphibians, and most reptiles, the X and the Y or the Z and the W of each species are still largely homologous to one other. Thus, even the heterogametic sex maintains two doses of each sex-linked gene. In the case of placental mammals, however, the Y has shed all the Mendelian genes which were allelic to the genes on the X. As a result, most, if not all, of the X-linked genes exist in the hemizygous state in the male. Each X-linked gene must have accommodated itself to this hemizygous state by doubling the rate of product output. Once this doubling in efficiency was accomplished, the genetic disparity between the male with one X and the female with two X's became very great. A need arose to adjust the dosage effect of X-linked genes between two sexes. In mammals, the dosage compensation is accomplished by random inactivation of one or the other X in individual female somatic cells. Consequently, phenotypic expression of X-linked genes in individual somatic cells of both sexes is hemizygous, and the mammalian female is a natural mosaic with regard to the activity of X-linked genes. It will be shown that the development of this particular form of dosage compensation mechanism is one of the reasons why the original X of a common ancestor was not fragmented into separate pieces during extensive speciation. Most of the X-autosome translocations disrupt the dosage compensation mechanism.

In avian species, the Z-linked genes also exist in the hemizygous state in the heterogametic female sex. Yet there is no sign of dosage compensation for these genes. On the contrary, the avian Z-linked genes show a definite dosage effect. The full manifestation of a Z-linked mutant phenotype requires the presence of two doses of a mutant gene in the homozygous male. The phenotype of a hemizygous female with a single dose of a mutant gene simulates that of the heterozygous male.

Part III attempts to elucidate the nature of sex-determining factors and the act of sex determination. When the X and the Y, or the Z and the W, were still largely homologous to one other, a male-determining factor on one, and a female-determining factor on the other chromosome, may have evolved as alleles of each other. During further differentiation, however, the sex-determining factors appear to have acquired certain peculiarities which set them apart from ordinary Mendelian genes. Although we now know that in placental mammals the Y-chromosome as a whole is very strongly male-determining, we have no answer to the important question of how many different kinds of male-determining factors there are on the mammalian Y. Circumstantial evidence shall be presented which suggests that there may be only a few different kinds of male-determining factors, but that they exist in multiplicates distributed along the entire length of the Y. The female-determining factors which presumably reside on the mammalian X shall also be considered. In the case of the Z and the W of avian species, we are totally ignorant of the nature of their sex-determining factors.

Many genes express themselves selectively through a particular cell type of the body at a particular time of ontogenic development. The structural gene for the γ-polypeptide chain of hemoglobin molecules, for instance, acts only through erythroid cells during fetal life. It will be shown that the sex-determining factors express themselves through somatic elements of the embryonic-indifferent gonads. When the male-determining factors prevail, interstitial cells, which are the producers of androgenic hormones develop, and the testis is organized. In the absence of competition from male-determining factors, the female-determining factors induce the development of follicular cells which are the producers of estrogenic hormones. The ovary results.

References

BALTZER, F.: Experiments on sex development in *Bonellia*. The Collecting Net 10, 3—8 (1935).
LEDERBERG, J.: Bacterial reproduction. Harvey Lectures Ser. **52**, 69—82 (1959).

Part I

Evolution of Vertebrate Sex Chromosomes and Genomes

Chapter 1

Evidence Indicating that the X and the Y or the Z and the W were Originally an Homologous Pair of Ordinary Chromosomes

Today, the X and Y of mammals and the Z and W of birds are totally different from each other in size as well as genetic content. Yet, it can be shown that the two were originally an homologous pair of ordinary chromosomes or autosomes. A broad review of the sex-determining mechanism of various vertebrates enables us to reconstruct, step-by-step, this process of sex chromosome differentiation.

a) The Absence of the Chromosomal Sex-determining Mechanism in Certain Teleost Fishes

As will be shown in Part III, the androgenic hormone-producing system (testicular interstitial cells) and the estrogenic hormone-producing system (ovarian follicular cells) share a common embryonic blastema. It may be surmised that in the total absence of the chromosomal (genetic) sex-determining mechanism, each individual vertebrate should develop into a functional hermaphrodite. Indeed, among the teleost fishes of today, synchronous hermaphroditism is a characteristic of the seabass belonging to the genera *Serranus* and *Hypoplectrus* of the family *Serranidae*, the order *Perciformes* (ATZ, 1964). This condition is also possessed by one species, *Rivulus marmoratus* of the family *Cyprinodontidae*, the order *Microcyprini* (HARRINGTON, 1963).

In synchronous hermaphroditism, the mature fish completes simultaneously both oögenesis and spermatogenesis. Two lobes of the ovotestis are posteriorly fused, their cavities joining to form a common oviduct. A greater part of each lobe is occupied by ovarian lamellae, and a small part by testicular tissue. Although a common sperm duct opens independently to the exterior, self-fertilization as a possibility in synchronous hermaphrodites has been confirmed under laboratory conditions with *Rivulus marmoratus* (HARRINGTON, 1963).

Functional hermaphroditism can also be seen in the protogynous and protoandrous forms. Among those belonging to the family *Serranidae*, the order *Perciformes*, all the groupers and their relatives, *Epinephelus*, *Mycteroperca*, *Alphestes*, *Petrometopon* and *Cephalopholis*, are protogynous. In the gonads of the protogynous species of the family *Serranidae*, ovarian lamellae fill the gonadal cavity during the female phase. In these lamellae, seminiferous crypts develop while the oöcytes degenerate as the male phase supercedes the female (SMITH, 1959). In one species, *Centropristes striatus*, the sex reversal from the female to the male was found to occur around the fifth year of life (LAVENDA, 1949).

The synchronous hermaphroditism described above requires a special isolation mechanism which prevents androgenic hormones of testicular tissue from exerting a suppressing influence upon neighboring ovarian tissue. Thus, it cannot be regarded as a most primitive state of sex differentiation, yet, such a state surely reflects the total lack of chromosomal sex-determining factors. The protogynous and protoandrous types of functional hermaphroditism, on the other hand, may be regarded as most primitive because here sex-determination is the result of the aging process.

At any rate, the persistence of functional hermaphroditism into certain teleost fishes of today is one evidence that, within the genome of ancestral vertebrates (which emerged for the first time 300 million years ago) there was no particular chromosome which had accumulated enough sex-determining factors to be qualified as the sex chromosome.

b) The Absence of Cytologically Detectable Sex Chromosomes in Lower Vertebrates

Functional hermaphroditism occurs only in certain member species of the orders *Perciformes*, *Microcyprini*, and *Myctophiformes* among the teleost fishes. A great majority of the teleosts are gonochorists, individual members of the species being either males or females. In amphibians and reptiles, hermaphroditism, functional or otherwise occurs only as an exception and never as a rule in a species.

In gonochoristic species, the first question to be asked is, "Which is the heterogametic sex?" In mammals and birds, the heterogametic sex can be determined by cytological means because the Y of the mammalian male and the W of the avian female are minute in size compared with the X and Z. In certain species, such as the rat (*Rattus norvegicus*, $2n = 42$), the autosomal constitution is such that the large X and the small Y cannot be distinguished at male mitotic metaphase. The first meiotic metaphase figure of the male, however, clearly reveals the presence of one asymmetrical bivalent made of one large X and one small Y.

In reptiles, amphibians, and fishes, it has long been MATTHEY's contention (1949) that the morphologically recognizable sex chromosomes do not exist. With the notable exception of snakes, which will be described later, our own experiences with lower vertebrates tend to support MATTHEY's contention. In many species of lower vertebrates, however, the diploid chromosome complements are such that the failure to recognize the heteromorphic chromosome pair at mitotic metaphase of either sex does not necessarily rule out the possibility that heteromorphic sex chromosomes might exist. Analysis of first meiotic metaphase figures from testes is of no help if the female is the heterogametic sex. In certain species of lower vertebrates, each component pair of the diploid complement can be recognized individually. In these species, the identical appearance of the male and the female diploid complements can be used as the positive proof that the X and Y or the Z and W of these species are morphologically identical. Figure 1 illustrates this point on the male and female karyotypes of the chameleon lizard (*Anolis carolinensis*, $2n = 36$) of the family *Iguanidae*. One sex has to be heterogametic, and one particular chromosome pair of that sex must represent either the XY- or the ZW-pair. Obviously, the X and Y or the Z and W are morphologically identical.

While not denying the possibility that there might be certain exceptional species of fishes and amphibians which have evolved the heteromorphic sex elements, the broad generalization can be made that in gonochoristic species of lower vertebrates the sex chromosomes are still in a primitive state of differentiation so that the X and Y or the Z and W of the heterogametic sex preserve identical morphological

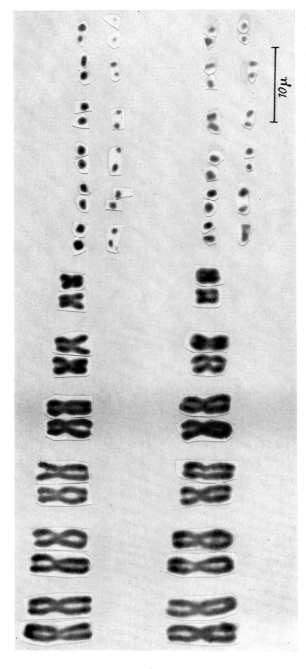

Fig. 1. The identical appearance of the male (top row) and female (bottom row) diploid complements of the chameleon lizard (*Anolis carolinensis*, 2n = 36)

appearance. Among lower vertebrates, the heteromorphic sex elements are the constant features of only certain families of snakes.

c) The Presence of the Undifferentiated Sex Chromosomes in Fishes and Amphibians as Revealed by Breeding Experiments with Sex-reversed Animals

In the absence of morphologically recognizable sex elements, other approaches had to be devised to elucidate the chromosomal sex-determining mechanism of lower vertebrates. In 1931, HUMPHREY pioneered one approach by achieving sex reversal through grafting an embryonic gonadal primordium of the genetic female to the male.

In *Ambystoma tigrinum* and *Ambystoma mexicanum,* a graft of ectoderm and underlying mesoderm was removed from an embryo of the tail-bud stage and was exchanged with a similar graft from another embryo of the same developmental stage. The ectoderm and mesoderm of the graft contained the primordia of the kidney and gonads which developed and differentiated in the new host. Thus, it was possible to obtain larvae having the ovary on one side of the body and the testis on the other. When the female graft was transplanted to the male, the host testis was often able to modify the grafted ovary. After the ovary was sufficiently modified to develop testicular lobules, the transformation to functioning testis continued after the removal of the host testis. In such a way, it became possible to produce a male who formed gametes from genetically female germ cells.

When such males were mated with the ordinary females, HUMPHREY accomplished the mating between the two genetic females. If the female is the homogametic sex (XX) in *Ambystoma,* all the offspring of this mating should be females. In reality, the mating produced offspring of both sexes with a preponderance of females, which can be expected if the female is the heterogametic sex of the ZW-sex-chromosome constitution. Three types of sex chromosome constitutions should appear among the offspring, with the ZZ : ZW : WW ratio of 1 : 2 : 1. Indeed, there were 25% males and 75% females. One-third of the females should be of the WW-constitution. When the WW-female obtained by this manner was mated with the ordinary male (ZZ), all the offspring were females (ZW). Thus, this experiment revealed the presence of the ZZ-ZW-type of the chromosomal sex-determining mechanism in *Ambystoma,* and the persistence of great homology between the Z and the W. Unless the W is a genetical equivalent of the Z, excluding a few opposing sex-determining factors, the WW-constitution would surely be a lethal formula (HUMPHREY, 1943).

In the South African clawed frog *(Xenopus laevis),* sex reversal was accomplished by adding estradiol to the water tank in which larvae were reared (GALLIEN, 1953; CHANG and WITSCHI, 1956). All treated larvae developed into functional females, but half of these had to be sex-reversed genetic males. Indeed, when mated to ordinary males, this half gave all male offspring: thus, the male of this species had to be the homogametic sex of ZZ-constitution.

Later, MIKAMO and WITSCHI (1964) were able to induce testicular development in genetic (ZW) female gonads of *Xenopus* by the same grafting technique used earlier by HUMPHREY on *Ambystoma.* When such sex-reversed genetic females were mated to ordinary ZW-females, functional WW-females were produced. As expected, the mating between the WW-female and the ordinary ZZ-male produced all female offspring. Furthermore, by the same grafting technique, it was possible to transform larvae of the WW-constitution into functional males. Thus, in the case of *Xenopus,* too, the Z of the heterogametic sex can be substituted by the W without any hindrance to the developmental process of individuals. Indeed, the Z and W differ only by a few opposing sex-determining factors.

In the Japanese cyprinodont fish Medaka *(Oryzias latipes),* the male heterogamety was proven by AIDA as early as 1921 by the use of allelic polymorphism of the partially sex-linked gene locus. The allele R (Ruby), normally situated on the Y, is dominant over the allele r (white) which is normally situated on the X, but crossing over between the X and the Y is effectively prevented (see Chapter 2). Thus, the mating can be set up where all male $(X^r Y^R)$ offspring are red, and all females $(X^r X^r)$, white. The stock would perpetuate this color difference betwen sexes.

Again by the use of estradiol, the sex reversal was accomplished in this species by YAMAMOTO (1961). When sex-reversed red females of the $X^r Y^R$-constitution were mated with ordinary males of the same sex chromosome constitution, one-third of the male offspring were expected to be of the $Y^R Y^R$-constitution. Although it was found that the $Y^R Y^R$-male is poorly viable, the cross-over $(Y^r Y^R)$ males were fully viable and fertile. When mated to ordinary white (X^rX^r) females, these cross-over males produced all male (X^rY^R and X^rY^r) offspring. Here again, the substitution of a single X of the heterogametic sex by a Y is compatible with the normal development of an individual. Figure 2 schematically illustrates the production of YY-males by sex reversal and the verification of the YY-constitution by the progeny test in the Medaka *(Oryzias latipes)*.

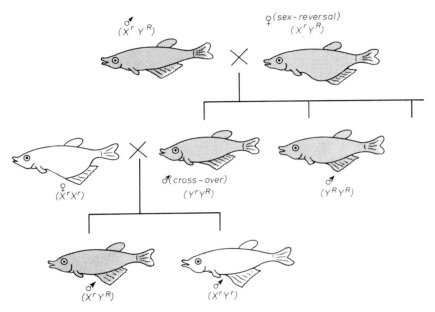

Fig. 2. A diagram illustrating the progeny of estradiol-sex-reversal female (X^rY^R) mated with normal (X^rY^R) male. The cross-over (Y^rY^R) male thus produced is fully viable and fertile. The Japanese cyprinodont fish *(Oryzias latipes)*

d) Interchangeability between the Male Heterogamety and the Female Heterogamety in the Fish

In species with primitive sex elements which differ from one another only by a few sex-determining gene loci, it is conceivable that the male heterogamety and the female heterogamety might have evolved simultaneously in different local races. In the moon-fish *(Xiphophorous maculatus)*, GORDON (1947) found that male heterogamety of the XY/XX-type operates in a local race inhabiting rivers of Mexico, while the aquarium stock and their wild ancestors inhabiting rivers in British Honduras operate on the female hetero-gamety of the ZZ/ZW-type (GORDON, 1951). The crossing of these two races revealed that the X of one race is a near equivalent of the W of the other. The same applies to the Y of one and the Z of the other.

e) Genetic Evidence of Homology between the X and Y or the Z and W of Lower Vertebrates

Another evidence which reveals a great homology between the X and the Y or the Z and the W of lower vertebrates is the fact that the genes on the X or the Z of fishes have their respective alleles on the Y or the W. As already mentioned, in the cyprinodont fish *Oryzias latipes,* the X-linked gene *r* (white) is a recessive allele of the gene *R* (ruby) which is normally on the Y (AIDA, 1921).

A similar situation was found in a number of partially sex-linked genes of the guppy, *Lebistes reticulatus* (WINGE, 1922; WINGE and DITLEVSEN, 1947). The sex-linked gene *elongatus* elongates the tail fin, and the gene *minutus* gives minute red spots on the base of the tail fin, while the gene *vitellinus* gives spots of egg-yolk color to the dorsal fin. Although their wild-type alleles are not readily identi-fiable, the very fact that these genes cross over from the X to the Y and from the Y to the X, reveals that both the X and the Y carry the gene loci for these alleles.

In an aquarium stock of the moon-fish *(Xiphophorous maculatus),* the ZZ/ZW-type of sex-determining mechanism operates (GORDON, 1951). The sex-linked gene *Dr* (red dorsal fin) and the *Rt* (ruby throat), as well as the gene *Sb* (spotted belly) and the *Sp* (spotted side) are allelic. In both cases, one allele may be carried by the Z and the other allele by the W in the heterogametic female.

References

AIDA, T.: On the inheritance of color in a fresh-water fish *Aplocheilus latipes* Temmick and Schlegel, with special reference to the sex-linked inheritance. Genetics 6, 554—573 (1921).

ATZ, J. W.: Intersexuality in fishes: In: Intersexuality in vertebrates includ-ing man, pp. 145—232. London and New York: Academic Press 1964.

CHANG, C. Y., and E. WITSCHI: Genic control and hormonal reversal of sex differentiation in *Xenopus*. Proc. Soc. exp. Biol. (N. Y.) **93**, 140—144 (1956).

GALLIEN, L.: Inversion totale du sexe chez *Xenopus laevis* Daud. á la suite d'un treatment gynogéne par le benzoate d'oestradiol administré pendant la vie larvaire. C. R. Acad. Sci. (Paris) **237**, 1565—1566 (1953).

GORDON, M.: Genetics of *Platypoecilus maculatus*. IV. The sex-determining mechanism in two wild populations of the Mexican platyfish. Genetics **32**, 8—17 (1947).

— Genetics of *Platypoecilus maculatus*. V. Heterogametic sex-determining mechanism in females of a domesticated stock originally from British Honduras. Zoologica **32**, 127—134 (1951).

HARRINGTON, R. W.: Twenty-four-hour rhythms of internal self-fertilization and of oviposition by hermaphrodites of *Rivulus marmoratus*. Physiol. Zool. **36**, 325—341 (1963).

HUMPHREY, R. R.: Sex inversion in the amphibia. Biol. Symp. **9**, 81—104 (1942).

LAVENDA, N.: Sexual differences and normal protogynous hermaphroditism in the Atlantic sea bass, *Centropristes striatus*. Copeia **3**, 185—194 (1949).

MATTHEY, R.: Les chromosomes de vertébrés. Lausanne 1949.

MIKAMO, K., and E. WITSCHI: Masculinization and breeding of the WW *Xenopus*. Experientia **20**, 622 (1964).

SMITH, C. L.: Hermaphroditism in some serranid fishes from Bermuda. Pap. Michigan Acad. Sci., Arts and Letters **44**, 111—119 (1959).

WINGE, O.: One-sided masculine and sex-linked inheritance in *Lebistes reticulatus*. J. Genetics **12**, 145—162 (1922).

—, and E. DITLEVSEN: Color inheritance and sex determination in *Lebistes*. Heredity **1**, 65—83 (1947).

YAMAMOTO, T.: Progenies of sex-reversal males in the medaka, *Oryzias latipes*. J. exp. Zool. **146**, 163—180 (1961).

Chapter 2

Differentiation of the Primitive Sex Elements at the Expense of the Y or W, and the Conservation of the Original X or Z

In the previous chapter it was shown that, with the exception of a small number of hermaphroditic species of fishes, the chromosomal sex-determining mechanism does exist in lower vertebrates. However, the X and the Y or the Z and the W of fishes and amphibians are still in an initial stage of differentiation. The two opposing sex determiners are still almost entirely homologous to each other.

In this chapter, an attempt shall be made to reconstruct the evolutional process of further sex chromosome differentiation which culminated in the grossly heteromorphic X and Y of mammals and the equally heteromorphic Z and W of birds.

It was our luck to find that among reptiles, the snake exhibits the Z and the W in three different stages of differentiation (BEÇAK et al., 1964). A great majority of snakes belonging to diverse families demonstrate the very similar diploid complements made of eight pairs of macrochromosomes and ten pairs of microchromosomes. The mediocentric chromosome which comprises about 10% of the genome (haploid set) constitutes the fourth largest pair of the homogametic male diploid set in each of these species.

As shown in Fig. 3, in the ancient family *Boidae* which includes such a giant as *Boa constrictor,* the fourth largest pair is homomorphic in both the male and the female. Thus, the Z and W of members of the *Boidae* are still in a primitive state of differentiation.

The family *Colubridae,* on the other hand, is a very prolific family reminiscent of rodents among mammals since it includes numerous species of diverse characteristics. In many members of this family, the Z and W differ by a pericentric inversion. The W is still as large as the Z; however, it is now a subterminal element. Poisonous snakes of the families *Crotalidae, Elapidae,* and *Viperidae* are regarded as most highly evolved of not only all the snakes, but also of all the reptiles. In members of the *Crotalidae,* the W of the female is as minute an element as the W of avian species, and a similarly minute W has been reported in members of the *Viperidae* by KOBEL (1962). Thus, the findings on snakes revealed two very pertinent facts with regard to the evolutional process of sex chromosome differentiation. First, the further differentiation from the initially homomorphic and largely homologous sex pair is accomplished exclusively at the expense of the element which is elected to accumulate factors governing the development of the heterogametic sex (the W in the case of the female heterogamety). On the contrary, the element which is elected to accumulate the determining factors of the homogametic sex (the Z in the female heterogamety) remained inviolate and is conserved in its entirety. Second, the first morphologically identifiable step of differentiation undertaken by the W is a pericentric inversion. These two pertinent facts which emerged from the study on snakes will be examined in greater detail.

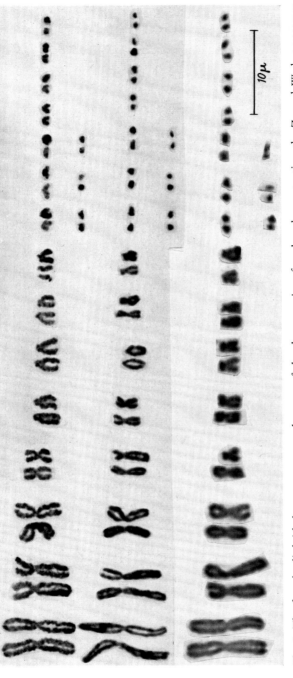

Fig. 3. The female diploid chromosome complements of the three species of snakes demonstrating the Z- and W-chromosomes in three different stages of differentiation. The fourth largest pair is the sex pair. Top row: The Z and the W are homomorphic in the South American epicrates (*Epicrates cenchria crassus*, 2n = 36) of the family *Boidae*. Middle row: The Z and the W differ by a pericentric inversion in the gopher snake (*Drymarchon corais couperi*, 2n = 36) of the family *Colubridae*. Bottom row: The W is much smaller than the Z in the side-winder rattlesnake (*Crotalus cerastes*, 2n = 36) of the family *Crotalidae*

a) Isolation during Meiosis of the Heterogametic Sex as an Essential Prerequisite of Sex Chromosome Differentiation

The differential accumulation of opposing sex-determining factors by two members of an originally homologous pair is possible only if the two remain isolated from each other during meiosis of the heterogametic sex. Free crossing-over would bring male and female determining factors differentially accumulated into the same chromosome, and chaos would result.

The very fact that such fishes as *Oryzias latipes, Lebistes reticulatus, Xiphophorus maculatus,* and such amphibians as *Ambystoma tigrinum* and *Xenopus laevis* do have a chromosomal sex-determining mechanism, reveals that the measure of isolation between the X and Y, or the Z and W, has been established in these species, despite the fact that the two opposing sex determiners are still almost entirely homologous to each other.

As described earlier, in the cyprinodont fish *Oryzias latipes* the dominant gene R (ruby), which is normally on the Y, is allelic to the X-linked gene r (white). Yet, the R on the Y is transposed to the X by crossing over at the extremely low frequency of 0.2%; therefore, the stock breeds true for red males (X^rY^R) and white females (X^rX^r). White males (X^rY^r) and red females (X^rX^R) are exceedingly rare. The segment involving this gene locus of the X is almost completely isolated from the Y in this species (AIDA, 1921).

It is of extreme interest to note that the effectiveness of this isolation mechanism in *Oryzias latipes* appears to depend not so much on the XY-constitution of germ cells, but rather on the testicular environment in which these germ cells are placed. When the genetic (X^rY^R) males are sex-reversed to the female and mated to the genetic (X^rX^r) female sex-reversed to the male, white (X^rY^r) males and red (X^rX^R) females are no longer rare among the offspring. The crossing-over between the X and Y occurs with five-times greater frequency when the XY-germ cells are placed in the ovarian environment (YAMAMOTO, 1961).

From the above it may be concluded that the initial isolation mechanism between the sex elements during meiosis of the heterogametic sex is furnished not by the sex elements themselves, but by the gonadal environment of that sex.

b) The So-called Differential and Homologous Segments of the Sex Chromosome

The concept that the sex chromosome is made of the differential and homologous segments is based on the assumption that this isolation involves only a part and not a whole of the X and Y or the Z and W.

In the guppy *(Lebistes reticulatus)*, the three non-allelic sex-linked genes, *elongatus, minutus,* and *vitellinus,* cross over from the X to the Y and from the Y to the X during male meiosis. On the contrary, the gene *coccineus* permanently remains on the X and the gene *sanguineus,* which gives a large red spot on the posterior-ventral portion of the dorsal fin, remains on the Y (WINGE and DITLEVSEN, 1947). On this basis, it was proposed that the first three genes, *el, mi,* and *vi,* are located on the homologous segment shared by the X and the Y. The gene *(co)* and the female-determining gene, on the other hand, are contained in the differential segment of the X. The gene *(sa)* and the male-determining gene are located in the differential segment of the Y.

On the surface, this concept of the homologous and differential segments is so attractive that an attempt has been made to extend this concept to cover the well-differentiated sex chromosomes of mammals (KOLLER and DARLINGTON, 1934).

Upon close examination, however, this concept suffers from two difficulties. First, the term, "the differential segment," implies that the isolation of this segment during heterogametic meiosis is a consequence of previous differentiation which resulted in the loss of homology between the X and Y at this segment. On the contrary, the isolation is an essential prerequisite of sex chromosome differentiation. The fact that the effective isolation is already established between the homologous part of the X and Y is well illustrated by the gene locus *(R, r)* of *Oryzias latipes* previously mentioned.

Second, the concept assumes that differential accumulation of opposing sex-determining factors occurs only in a small segment of the X and Y, or the Z and W. It should be realized that if only a single gene locus is involved in sex determination as visualized by WINGE on *Lebistes reticulatus,* there is absolutely no need for the isolation between the X and Y.

WINGE felt that in *Lebistes reticulatus* the male-determining gene (M) is a dominant allele of the female-determining gene (f). Thus, the homozygous state (f, f) is required for the femaleness, while the heterozygous state (M, f) is sufficient for the maleness. If the sex is determined by a pair of alleles, unrestricted crossing-over between the X^f and the Y^M during male meiosis in no way hinders the sex-determining mechanism. The mating between the f/f-genotype (female) and the M/f-genotype (male) invariably results in the production of the two parental genotypes in the exact ratio of one-to-one.

The isolation is meaningful only because it permits a relatively long segment of the X as well as the Y to gradually accumulate a number of sex-determining factors well spaced from each other. Thus it facilitates further differentiation of the primitive sex elements which are still largely homologous to each other.

The so-called differential segment is nothing but a part of the sex chromosome where, at a given stage of the evolutional process, the particularly effective isolation mechanism is operating.

c) Changes on the Y or the W as a Cause of the Effective Isolation

It has been established that in *Oryzias latipes*, effectiveness of the isolation mechanism is influenced by a change in the gonadal environment. It is likely that as differentiation of the sex elements progressed, a firmer isolation mechanism evolved.

Although the brine shrimp *(Artemia salina)* is not a vertebrate, the results obtained by BOWEN (1965) on this species are so revealing that they are presented here. In this species, the female heterogamety of the ZZ/ZW-type is in operation. In wild populations of brine shrimp, the eye color is black. It can be shown that a recessive allele *w*, which governs white eyes, is on the sex chromosome. In matings of black-eyed (Z^wW^+) females to white-eyed (Z^wZ^w) males, cross-overs can be recognized among offspring as white-eyed (Z^wW^w) females and black-eyed (Z^wZ^+) males.

In the laboratory stock maintained by BOWEN, the frequency of crossing-over involving this gene locus was very low (0.06%). When crosses were made between this stock and seven wild races from Argentina, Mexico, Canada, and four localities in the United States, it was found that the W^+-chromosomes of wild races, when paired with a standard Z^w of this stock, demonstrated great variation in the frequency of crossing-over ranging from a low of 0.03% to a high of 20%. Further progeny tests revealed that each line descending from a single F_1-female had its own characteristic frequency of crossing over which was transmitted matroclinously. BOWEN was able to identify at least three types of W-chromosomes which have different frequencies of crossing-over: W_9, 0.03%; W_{13}, 1.3%; and W_{12}, 12%. Thus, it becomes evident that the constitution of the W-chromosome, the determiner of the heterogametic sex, decides the effectiveness in suppressing crossing-over between the Z and the W during the heterogametic meiosis. The Z-chromosome, the determiner of the homogametic sex, apparently plays no active role in the isolation mechanism. BOWEN suspects that inversions, deletions, or suppressor genes in the W-chromosomes are responsible for the isolation mechanism.

d) A Pericentric Inversion on the Y or the W as a First Step toward the Heteromorphism of the Sex Chromosomes

These findings on the brine shrimp (*Artemia salina*) revealed that a structural change which occurred on the W-chromosome might have been responsible in making the isolation of the Z from the W more stable.

This is in good agreement with our cytological findings on various snake species (BEÇAK et al., 1964). As shown in Fig. 3, a majority of the species belonging to the family *Colubridae* demonstrate the intermediate stage of sex chromosome differentiation. Although the Z and W of the female are still identical in size, a pericentric inversion has occurred in the W. While the Z remains a mediocentric element, the W now has become a subterminal element.

There is no reason why a pericentric inversion, *per se*, should suppress crossing-over between the Z and the W, but it does create a stronger need for the establishment of the permanent isolation mechanism. If crossing-over were permitted to occur in the inverted segment, two unbalanced chromosomes would arise, one partially duplicated and the other partially deleted.

By creating a need for permanent isolation, a pericentric inversion in the Y or the W facilitates further differentiation of sex chromosomes.

e) The Final Emergence of the Heteromorphic Sex Elements

The Z-chromosome of various ophidian as well as avian species comprises 10% of the genome (haploid set). These genes, which have been on the Z, must comprise an indispensable part of the genome. Their loss from the genome would have disastrous effects. This appears to be the very reason why the differentiation from the primitive homomorphic sex pair to the grossly heteromorphic ZW-pair is accomplished exclusively at the expense of the W, while the Z is conserved in its entirety (Fig. 3). Even after the W-chromosome differentiated into a minute element, the heterogametic female sex is still endowed with the Z which carries a full 10% load of the genome. The only thing the Z-linked genes have to do is to accommodate themselves to the hemizygous state in the female.

Once the nearly complete isolation mechanism was established between the Z and W which were still largely homologous to each other, gradual genetic deterioration of the W probably followed. The W-linked genes which had once been the alleles of the Z-linked genes must have deteriorated one by one, giving enough time for each Z-linked gene to accommodate itself to the hemizygous existence.

Once genetic deterioration was completed, the W became genetically empty except for the determiners of the heterogametic female sex it acquired during differentiation. Now the stage was set for the final emergence of the grossly heteromorphic Z and W. The above should apply equally well to the X and the Y of the male heterogamety.

Among the snakes of the family *Colubridae*, the inverted W of certain species might represent the stage where the deterioration of the W-linked genes is far enough advanced. It is tempting to think that the minute W of the family *Crotalidae* was produced in one sweep from the inverted W of these species. The fact that the snake W changed from a mediocentric element of the *Boidae* to a subterminal element of the *Colubridae* reveals that the segment involved in a pericentric inversion was almost as large as one entire arm of the mediocentric. If there was an accidental breakdown of the isolation mechanism, crossing-over between the Z and an inverted segment of the W would result in the formation of the two new chromosomes. One of the new chromosomes would be made almost entirely of an inverted segment, and therefore be only half as large as the original W. If a crossing-over occurred near the end of an inverted segment, the Z contributed very little of its material to this new chromosome. It may be that the new chromosome produced in this manner is the direct precursor of the minute W which we see in poisonous snakes of the families *Crotalidae*, *Viperidae*, and *Elapidae*.

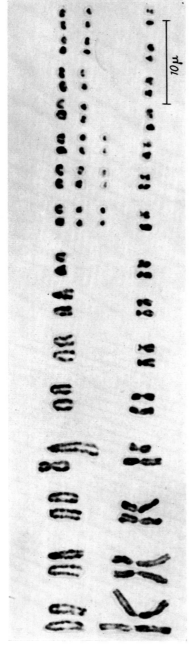

Fig. 4. The female diploid complements of the two exceptional species of snakes belonging to the family *Colubridae*. The fourth largest pair is again the sex pair. Top row: The South American clelia (*Clelia occipitolutea*, 2n = 50) has the W-chromosome nearly twice the size of the Z. Bottom row: The South American xenodon (*Xenodon merremii*, 2n = 30) has the W which is as minute as those possessed by members of the family *Crotalidae*

It is conceivable that the deterioration of Mendelian genes was accompanied by redundant multiplication of each kind of sex-determining factor along the entire length of the inverted W. The elimination by an accidental crossing-over of half the chromosome material from such a W is the next logical step of sex chromosome differentiation.

The other viable product of an accidental crossing-over between the inverted W and the Z is a new W-chromosome which is much larger than the Z, for it incorporated nearly one entire arm of the Z to the inverted W.

Among South American members of the *Colubridae*, one exceptional species *(Xenodon merremii*, 2n = 30) demonstrated the W which was as minute as the W of poisonous snakes of the family *Crotalidae*. The other exceptional species *(Clelia occipitolutea*, 2n = 50) had the W which greatly exceeded the Z in size (Fig. 4). The presence of both the minute W and the very large W within the family *Colubridae* lends a measure of creditability to the notion that the final emergence of the truly specialized W is because of an accidental crossing-over between the Z and the pericentrically inverted W.

Additional support for this notion is furnished by the fact that the presence of the W-chromosome considerably larger than the Z has been reported in two other species already mentioned in this book. In both species, genetic evidence indicates that the Z and W are still in a primitive state of differentiation. In the African clawed frog *(Xenopus laevis*, 2n = 36), females of the stock maintained by a commercial breeder in Southern California exhibited one very large mediocentric chromosome which was not seen in males of the same stock. This element was interpreted to be the W (WEILER and OHNO, 1962). Similarly, in the brine shrimp *(Artemia salina)*, STEFANI (1963) found that in half of the blastulae of the bisexual population at San Bartoloneo, Italy, a conspicuously long chromosome existed alone, while in the other half of the blastulae, the diploid complement was made of the 21 homologous pairs of small chromosomes. When the parthenogenic population at San Gilla was studied, the same long chromosome was found in all the blastulae. Accordingly, this chromosome was interpreted as being the W of the female.

In the phylogeny of sex chromosome differentiation, such large W-chromosomes appear to represent a step down a dead-end street. Nevertheless, the sporadic occurrence of the W which is larger than the Z among species with relatively undifferentiated sex chromosomes seems to suggest that accidental crossing-over between the Z and the inverted W does occur. If this occurs to the inverted W which has not undergone sufficient genetic deterioration, the species may not be ready to tolerate the minute W which lost so much chromosome material, and such a species would maintain only the partially duplicated product (the larger W).

The entire process of sex chromosome differentiation from a homologous pair of ordinary chromosomes is thus reconstructed step by step. The process is schematically illustrated in Fig. 5.

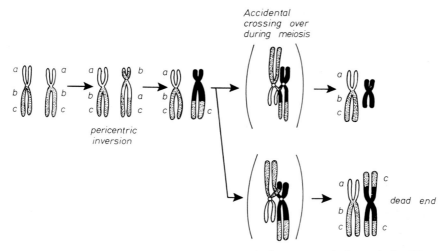

Fig. 5. A schematic representation of the phylogenic evolution of the X and the Y or the Z and the W from a primitive homomorphic pair (extreme left) to a grossly heteromorphic pair (extreme upper right). *a*, *b*, *c* denote the Mendelian genes which were originally there when the X and the Y or the Z and the W were merely a homologous pair of ordinary chromosomes. An area of the Y or the W where genetic deterioration occurred is indicated by solid black

References

AIDA, T.: On the inheritance of color in a fresh-water fish *Aplocheilus latipes* Temmick and Schlegel, with special reference to the sex-linked inheritance. Genetics 6, 554—573 (1921).

BEÇAK, W., M. L. BEÇAK, H. R. S. NAZARETH, and S. OHNO: Close karyological kinship between the reptilian suborder *Serpentes* and the class *Aves*. Chromosoma (Berl.) **15**, 606—617 (1964).

BOWEN, S. T.: The genetics of *Artemia salina*. V. Crossing-over between the X and Y chromosomes. Genetics **52**, 695—710 (1965).

KOBEL, H. R.: Heterochromosomen bei *Vipera berus* L. *(Viperidae, Serpentes)*. Experientia **18**, 173—174 (1962).

KOLLER, P. C., and C. D. DARLINGTON: The genetical and mechanical properties of the sex chromosomes. I. *Rattus norvegicus*. J. Genet. **29**, 159 (1934).

STEFANI, R.: La digametia femminile in *Artemia salina* Leach e la costituzione del corredo cromosomico nei biotipi diploide anfigonico e diploide partenogeneico. Caryologia **16**, 625—636 (1963).

WEILER, C., and S. OHNO: Cytological confirmation of female heterogamety in the African water frog *(Xenopus laevis)*. Cytogenetics **1**, 217 —223 (1962).

WINGE, O., and E. DITLEVSEN: Color inheritance and sex determination in *Lebistes*. Heredity **1**, 65—83 (1947).

YAMAMOTO, T.: Progenies of sex-reversal females mated with sex-reversal males in the medaka, *Oryzias latipes*. J. exp. Zool. **146**, 163—180 (1961).

Chapter 3

Polyphyletic Evolution of Vertebrates

Thus far we have seen that, during the course of sex chromosome differentiation, the one which was elected to accumulate factors governing the development of the homogametic sex was conserved in its entirety. The differentiation was accomplished exclusively at the expense of the element which was elected to be the determiner of the heterogametic sex, the Y or the W.

This conservation of the X or the Z can be readily understood if it is realized that the genes on the predecessor of the sex element must have constituted an indispensable part of the genome. Thus, while genetic deterioration was permitted to occur to the Y or the W, all these genes had to be maintained by the other member of the sex pair. It then follows that various species which descended from an immediate common ancestor should share the same class of sex-linked genes.

It is known that all the placental mammals of today descended from a common ancestor, protoinsectivores, which emerged at the dawn of the cenozoic era, 70—135 million years ago. Before the homology of X-linked genes is sought among placental mammals, however, it will have to be established that the total genome of the common ancestor has not changed substantially in quantity during extensive speciation that followed. The conservation of the X-chromosome is meaningful only within the context of the stable genome.

Similarly, the search for the homology of Z-linked genes among avian species becomes meaningful only if extensive speciation within the class *Aves* was accomplished with little or no change in the total genetic content.

The history of vertebrates began about 300 million years ago in the Devonian period of the paleozoic era. The route of evolution from aquatic to terrestrial forms, and then to the final emergence of the class *Aves* on one hand and the class *Mammalia* on the other had not been a single straight path. As will be shown, gene duplication appears to have played a major role in the evolution of vertebrates. Surviving members of the classes *Pisces, Amphibia,* and *Reptilia* are likely to be the representatives of several diverse evolutionary pathways, with each genome lineage containing a characteristic degree of gene duplication. The genomes of only some of these fishes, amphibians, and reptiles are expected to be directly related to the genomes of mammals and birds. It seems desirable to obtain a comprehensive picture of the evolution of vertebrate genomes before we pursue the topics of the conservation of the original X by placental mammals and of the original Z by various avian species.

a) Importance of Gene Duplication in Vertebrate Evolution

Needless to say, speciation is the consequence of genetic changes accumulated within the genome. In the case of speciation from an immediate ancestor, genetic changes are no doubt due mainly to allelic mutations. What used to be a rare mutant allele of the old species would come to exist as the wild-type allele in a new species.

When the scope is broadened to consider the evolution of vertebrates as a whole, allelic mutations of already existing genes cannot possibly account for all the genetic changes that occurred during 300 million years. Gene duplication now emerges as the single most important factor in evolution.

The extensive study carried out by MARGOLIASH (1963) on molecular structures of cytochrome c revealed the extremely conservative nature of the gene. Cytochrome c is the heme-containing protein which engages in the intracellular transportation of oxygen. As such, it must have come into existence soon after cells made their first appearance on this earth as the unit of life. Yet, it was found that cytochrome c of diverse organisms, from yeasts to man, not only have nearly the same molecular weight, but also maintain similar amino acid sequences, with about 104 amino acid residues making up a polypeptide chain in each instance. The clear implication is that the particular function assigned to the gene product imposes a severe limitation on that gene's freedom to mutate. If a change in the base sequence occurred at the wrong part of the DNA molecule, a new gene product would be unable to function as cytochrome c. Such a mutation would quickly be eliminated.

Natural selection conserved only those mutations which were not deleterious to the gene product's assigned function. The extremely conservative nature of the already existing genes indicates to us that the redundancy of genetic material was the prerequisite for the creation of new genes. By duplication, if the old gene had been represented twice within the genome, one of the duplicates was now free to mutate to an independent direction and acquire a new function.

In man and probably most other mammals, there are five independent gene loci for component polypeptides of hemoglobin. They are for α-, ε-, γ-, β-, and δ-chains.

It is the view of INGRAM (1963) that the ancestry of all the five genes for five different components of hemoglobin can be traced back to a single ancestral gene. First, there was a duplication of this gene, and by subsequent mutations in independent directions, one became a gene for myoglobin while the other became a gene for α-chain of hemoglobin. When it first emerged, the ancient vertebrate may have been able to produce only one type of a hemoglobin molecule which should be α_4. The genes for four other chains of hemoglobin are thought to have been derived from the multiplicates of the gene for the α-chain.

Similarly, mammals and birds have three independent gene loci for component polypeptides of the enzyme, lactate dehydrogenase. They are known as A, B, and C (MARKERT, 1964; BLANCO and ZINKHAM, 1962; BLANCO et al., 1964). Originally there may have been only one gene locus for LDH, and the other two may have been produced by duplication of the original one.

Gene duplication can be accomplished in two different ways. The longitudinal duplication of a small segment of an individual chromosome would accomplish the purpose for a small number of genes closely linked together. In fact, regional duplication of small chromosomal segments appears to be occurring among mammals today. For instance, the γ-globulin molecule is made up of two different kinds of polypeptide chains: the heavy-chain (H) with a molecular weight of about 60,000 and the light-chain (L) of about 20,000. In man, it is becoming increasingly clear that, instead of having one gene locus each for the H- and L-chains, the so-called H-chain locus is actually made of several very closely linked, but slightly different genes. The same can be said of the so-called L-chain locus. There apparently was longitudinal multiplication of one ancestral gene for the H-chain and the other ancestral gene for the L-chain. In man, the genes for β- and δ-chains of hemoglobin are also very closely linked. The δ-chain gene must have been derived by a regional duplication of the β-chain gene.

While regional duplication of a small number of genes might have played an important role in speciation from an immediate ancestor, more drastic changes must have occurred to the genomes during the course of vertebrate evolution. Simultaneous duplication of the entire set of genes can be accomplished by polyploidization. It can be assumed with reasonable certainty that a series of polyploidization of the ancestral genome has taken place sometime in the history of vertebrates, which is, after all, 300 million years old.

b) Incompatibility between Polyploidy and the Well Established Chromosomal Sex-determining Mechanism

Polyploidy is incompatible with the well established chromosomal sex-determining mechanism. When diploid organisms with the XY/XX-scheme of sex-determining mechanism become tetraploid, the male has to maintain the 4AXXYY-constitution and the female, 4AXXXX. During meiosis of the 4AXXYY-male, the four sex elements may pair off as the XX-bivalent and the YY-bivalent. If this occurs, every gamete would be of the 2AXY-constitution. Consequently, all the offspring resulting from the mating between a tetraploid male and a tetraploid female would emerge as the intersex of the 4AXXXY-constitution. Even if two XY-bivalents are formed in individual spermatocytes of the tetraploid male, in 50 per cent of the cases the X and the Y would move to the same division pole at first meiotic anaphase, again resulting in the production of the inter-sex of the 4AXXXY-constitution. Polyploidy invariably disturbs the chromosomal sex-determining mechanism.

Indeed, in vertebrates, viable and fertile polyploid individuals have been found among amphibians where the Z and the W are still largely homologous to each other. The same is not true in birds and mammals where the W and the Y became a highly specialized deter-miner of the heterogametic sex. Furthermore, it was found that even in amphibians, polyploid individuals are, as a rule, incapable of perpetuating themselves as polyploid by bisexual mating.

Polyploid individuals of amphibians were most thoroughly studied by HUMPHREY and his colleague (HUMPHREY and FANKHAUSER, 1956; FANKHAUSER and HUMPHREY, 1959) on *Ambystoma mexicanum* and *A. tigrinum*.

In the case of triploidy, males were uniformly of the 3AZZZ-constitution, while three kinds of sex chromosome constitutions, ZZW, ZWW, and WWW were found among females. Triploids of both sexes were of very poor fertility, and males were more sterile than females; therefore, it was not possible to perpetuate the triploid race by mating of triploid males and triploid females. When mated to diploid males, triploid females produced many tetraploids, revealing that these females ovulate triploid eggs. This may account for the presence of a gynogenic all-female triploid race found in *Ambystoma jeffersonianum* (UZZELL, 1963). Tetraploid, pentaploid, hexaploid, and heptaploid individuals of *A. mexicanum* and *A. tigrinum* also showed very poor fertility. In short, it appears that even in amphib-ians with the undifferentiated sex chromosomes, a serious obstacle which prevents the emergence of a bisexual polyploid race exists.

From the above, it may be deduced that various degrees of poly-ploidization of the ancestral genome must have occurred very early in the evolution of vertebrates before the emergence of terrestrial forms.

Our study on DNA-contents of various vertebrates confirms the above prediction and reveals the polyphyletic origin of the genomes of various vertebrates.

c) Uniformity of the DNA Content of Various Placental Mammals

All placental mammals of today descended from a common stock of protoinsectivores which emerged at the dawn of the Cenozoic era. In terms of geological time, the history of placental mammals is brief indeed. Reflecting this recent origin is the sameness of DNA content. Diverse speciation appeared to be accomplished with little or no change in the total genetic content. MANDEL and his colleagues (1950) were among the first to show that each diploid nucleus of man as well as cattle, sheep, pigs, and dogs contains about 7.0×10^{-9} mg of DNA, no more, no less. Recently, we re-examined this matter of DNA constancy by means of microspectrophotometry. Six species representing four different orders were chosen: man *(Homo sapiens,* 2n = 46) representing the order *Primates,* the dog *(Canis familiaris,* 2n = 78) representing the order *Carnivora,* the horse *(Equus caballus,* 2n = 64) of the order *Perissodactyla,* the mouse *(Mus musculus,* 2n = 40), the golden hamster *(Mesocricetus auratus,* 2n = 44), and the creeping vole *(Microtus oregoni,* 2n = 17/18) of the order *Rodentia.* There was no significant difference in DNA values between man, the horse, the dog, the golden hamster, and the mouse. A single exception was the creeping vole which had a DNA value 10% lower. This species (OHNO et al., 1963) shares with the other member of the rodent subfamily *Microtinae, Ellobius lutescens,* 2n = 17 (MATTHEY, 1953) the distinction of having the lowest diploid chromosome number known among placental mammals. Such a drastic reduction in the number of chromosomes had to be accompanied by the loss of a number of centromeres with their adjacent heterochromatic chromosomal materials. This loss of genetically unimportant heterochromatin would account for the 10% lower DNA value found in the creeping vole (ATKIN et al., 1965).

It then follows that different species of mammals, by and large, share the same kinds of gene loci, even if they belong to different orders. Allelic mutations at each gene locus were mainly responsible for extensive diversification of placental mammals. The sameness of total genetic content, however, does not exclude the possibility that duplications of a small number of genes may have occurred during speciation of mammals.

Regional gene duplications which occurred to a small number of genes of different mammals, however, do not change the over-all

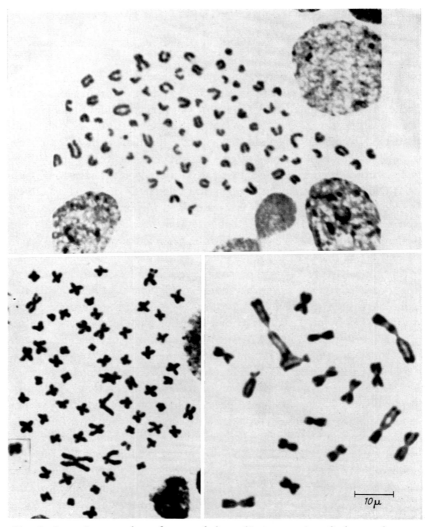

Fig. 6. Somatic metaphase figures of three diverse species of placental mammals printed at the same magnification. Despite marked differences in diploid chromosome number and morphology of individual chromosomes, the DNA values are about the same. Top: The male dog (*Canis familiaris*, 2n = 78). All except the X-chromosome (11 O'clock) are acrocentrics. Bottom left: The male chinchilla (*Chinchilla laniger*, 2n = 64). All the 64 chromosomes are metacentrics. The mediocentric X is the largest element. The X of this species is the duplicate-type (see Chapter IX). Bottom right: The female creeping vole (*Microtus oregoni*, 2n = 17). The X is the second largest element. It is the triplicate-type. The female soma of this species is normally XO (see Chapters IX and XI)

picture of mammalian evolution. Extensive speciation of placental mammals was accomplished without substantial change in the total genetic content. Yet, placental mammals of today display chromosome constitutions of infinite variety. The diploid number ranges from a high of 80 in the primitive primate, *Tarsius bancanus* (KLINGER, 1963) to a low of 17 in two rodent species mentioned above. Nothing but acrocentric chromosomes are found in the mouse *(Mus musculus*, 2n = 40), and only metacentrics in the chinchilla *(Chinchilla laniger*, 2n = 64) (GALTON et al., 1965). The enormous array of karyotypes reveals the extent to which the original autosomal linkage groups of a common ancestor have been shuffled around. An autosomal equivalent to the human chromosome 21 may only be found among his closest relatives, the chimpanzee *(Pan troglodytes*, 2n = 48) and the gorilla *(Gorilla gorilla*, 2n = 48) (HAMMERTON et al., 1963).

While the autosomes have broken and reunited many times, the X-chromosome has apparently persisted without substantial change from its beginning. This conservation of the entire ancestral X by various placental mammals of today will be discussed in great detail later in connection with the homology of X-linked genes.

Figure 6 demonstrates three extreme types of diploid complements of placental mammals. They are of the dog *(Canis familiaris*, 2n = 78), of the chinchilla *(Chinchilla laniger*, 2n = 64), and of the creeping vole *(Microtus oregoni*, 2n = 17/18).

d) Uniformity of the DNA Content of Various Avian Species

It is believed that ancestral forms of modern birds were already in existence near the end of the Jurassic period of the Mesozoic era. The fossil remains of the toothed bird *Archeopteryx lithographia* found in slate deposits in Bavaria, are said to be 150 million years old. Thus, it is clear that the avian lineage branched out from a reptilian lineage before the other reptilian lineage gave rise to a common ancestor of placental mammals.

Reflecting this independent evolution of the two classes of warm-blooded vertebrates is the fact that the male is the heterogametic sex in mammals, while in birds it is the female. The avian chromosome complements are also distinct from those of placental mammals in that they include numerous microchromosomes, each no larger than one micron.

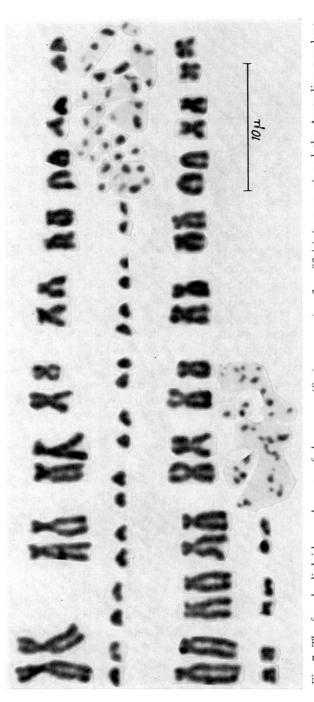

Fig. 7. The female diploid complements of the canary (*Serinus canarius*, 2n = 80 ±) (top rows) and the Australian parakeet (*Melopsittacus undulatus*, 2n = 60 ±) (bottom rows). The heteromorphic ZW pair constitutes the fourth largest in the canary and the fifth largest in the parakeet

In our experience, the diploid complements of present day birds belonging to the orders *Passeriformes, Columbiformes, Galliformes,* and *Anseriformes,* follow the common rule in that nine pairs of macrochromosomes, or ordinary chromosomes, and about 60 microchromosomes constitute each diploid complement. Members of the order *Psittaciformes* were exceptional, having more macrochromosomes and fewer microchromosomes. For instance, twelve pairs of macrochromosomes and about 18 pairs of microchromosomes constitute the diploid complement of the Australian parakeet, *Melopsittacus undulatus* (OHNO et al., 1964). The karyotypes of the canary *(Serinus canarius)* and the parakeet *(Melopsittacus undulatus)* are shown in Figure 7 as representatives of avian species.

Relative DNA values were measured on the canary representing the order *Passeriformes,* the chicken *(Gallus gallus domesticus)* representing the order *Galliformes,* the pigeon *(Columbia livia domestica)* of the order *Columbiformes,* and the Australian parakeet of the order *Psittaciformes.* Since the extreme similarity in their diploid chromosome complements indicates the uniformity in the total genetic content of various avian species, these four species representing four diverse orders were deemed sufficient.

As expected, four representatives of the class *Aves* gave the uniform DNA value. The value, however, was 44—59% that of placental mammals.

The above finding on the total genetic content of avian species, on one hand, reveals the fact that polyploidy played no role in extensive speciation within the class *Aves* and, on the other hand, shows that the genome lineage which gave rise to the class *Aves* has long been separated from that which eventually gave rise to placental mammals.

e) The Coexistence of the Two Genome Lineages in the Class Reptilia

Reptiles of today can be compared with the twigs of a great tree which flourished during the early Mesozoic era. About 95% of all the different kinds of living reptiles belong to the order *Squamata,* yet fossil remains indicate that this order never held greater importance than today. On the contrary, fossil beds in many parts of the world are strewn with shells of many kinds of turtles. The orders *Crocodylia* and *Chelonia* have seen better days.

We have no way of directly assessing the genomes of ancient reptiles which constituted the huge limbs of a great tree and which produced the direct ancestor of placental mammals on one hand and of birds on the other. It was fortunate that studies on the relative DNA contents and chromosome constitutions of a limited number of living reptiles enabled us to discern the presence of two different genome lineages, one showing close affinity to that of the class *Aves* and the other to that of the class *Mammalia*.

As stated earlier, the presence of abundant microchromosomes characterizes the avian chromosome complements. It has been known for some time that microchromosomes are possessed also by lizards and snakes which constitute the order *Squamata*. While the exact number of microchromosomes in the diploid complement of each avian species is nearly impossible to determine, the number of microchromosomes in each lizard or snake species can be determined with ease. The somatic metaphase figure of an alligator lizard *(Gerrhonotus multicarinatus,* 2n = 46—48) belonging to the family *Anguidae* of the suborder *Sauria,* contains 12 pairs of microchromosomes, as shown in Figure 8a. The fact that a majority of snake species constituting the suborder *Serpentes* contain 10 pairs of microchromosomes, was already mentioned in a previous chapter (Fig. 3).

Aside from the possession of microchromosomes, there is yet another common characteristic which reveals that the reptilian order *Squamata,* suborder *Serpentes* in particular, belong to the very same genome lineage which gave rise to the class *Aves*. The female heterogamety of the ZZ/ZW-type also operates in snakes (BEÇAK et al., 1962; KOBEL, 1962). Furthermore, the avian Z-chromosome and the ophidian Z-chromosome may have been derived from the same ancestral chromosome, as both constitute about 10% of the genome or haploid set (BEÇAK et al., 1964).

DNA content was measured in six representatives of the order *Squamata* which were: the chameleon lizard *(Anolis carolinensis,* 2n = 36) of the family *Iguanidae* and the alligator lizard *(Gerrhonotus multicarinatus,* 2n = 46) of the family *Anguidae* of the suborder *Sauria*. The suborder *Serpentes* was represented by the boa constrictor *(Boa constrictor amarali,* 2n = 36) of the family *Boidae,* the gopher snake *(Drymarchon corais couperi,* 2n = 36), and the South American Xenodon *(Xenodon merremii,* 2n = 30) of the family *Colubridae,* and the South American jararaca *(Bothrops jararaca,* 2n = 36) of the

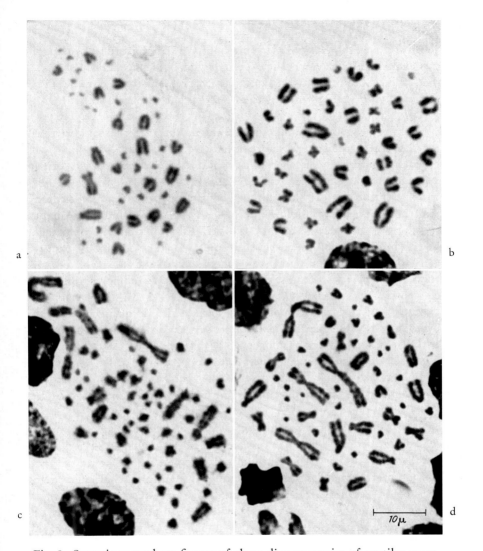

Fig. 8. Somatic metaphase figures of three diverse species of reptiles representing the three different orders. a) The alligator lizard (*Gerrhonotus multicarinatus*, 2n = 47) representing the order *Squamata*. The DNA value is
60% that of placental mammals. This individual was heterozygous for a
single Robertsonian translocation. b) The South American alligator (*Caiman
sclerops*, 2n = 42) representing the order *Crocodylia*. The DNA value is
84% that of placental mammals. c) and d) The fresh-water soft-shelled
turtle (*Amyda ferox*, 2n = 66) and the desert tortoise (*Gopherus agassizi*,
2n = 52) representing the order *Chelonia*. The DNA values are 80 and 89%
that of placental mammals, respectively

family *Crotalidae*. These six representatives of the order *Squamata* possessed a DNA value of 60—67% that of placental mammals. The value obtained was only slightly more than that obtained for various avian species which was 44—59% that of placental mammals (ATKIN et al., 1965).

While the above finding should not be interpreted to mean that lizards and snakes of today were directly ancestral to birds, it reveals that an ancestral reptile which evolved to toothed birds, belonged to the same genome lineage which independently gave rise to ancestral forms of modern members of the order *Squamata;* there was no further polyploidization of this genome lineage. Among members of the order *Squamata*, the well differentiated heteromorphic Z- and W-chromosomes are seen only in the poisonous family *Crotalidae* and certain members of the family *Colubridae* of the suborder *Serpentes;* others possess the primitive homomorphic sex elements. Yet apparently an effective barrier exists to prevent the evolution of a bisexual polyploid species. Triploid species of the Teiid lizard of the genus *Cnemidophorus* were all females and apparently propagated by parthenogenesis (PENNOCK, 1965).

While present day members of the reptilian order *Squamata* demonstrate the close kinship to the class *Aves*, members of the orders *Crocodylia* and *Chelonia* appear to represent the premammalian genome lineage.

The South American alligator *(Caiman sclerops,* 2n = 42) representing the order *Crocodylia* revealed the DNA value as 84% that of placental mammals. The diploid chromosome complement of this species shown in Figure 8b is totally different in character from that of snakes and lizards illustrated in Figures 3 and 8a. In fact, there is a striking resemblance between the diploid complement of *Caiman* and that of one species of mammals, the rat *(Rattus norvegicus,* 2n = 42). To be sure, this extreme similarity is a pure coincidence. Nevertheless, there is little doubt that among present day reptiles, those belonging to the order *Crocodylia* demonstrate the closest kinship with placental mammals, not only in DNA value, but in karyological characteristics as well. Although the lower diploid chromosome number of 32 has been reported on the North American alligator *(Alligator mississippiensis)* and the African crocodile *(Crocodilus niloticus)*, this reduction in chromosome number from 42 to 32 appears to be the result of simple Robertsonian translocations.

The 10 largest pairs of acrocentrics of *Caiman* are represented as 5 pairs of metacentrics in *Alligator* and *Crocodilus* (HOLLINGSWORTH, 1957; VAN BRINK, 1959).

A DNA value similar to that of placental mammals was also obtained in representatives of the order *Chelonia*. The fresh-water soft-shell turtle *(Amyda ferox,* 2n = 66) and the desert tortoise *(Gopherus agassizi,* 2n = 52) gave DNA values 80 and 89% that of placental mammals, respectively. Observing Figures 8c and 8d, however, it may be noted that their karyological characteristics are not at all similar to those of placental mammals. Many small members can be regarded as microchromosomes. It appears that members of the order *Chelonia* demonstrate the closest karyological affinity to the infraclass *Prototheria*, rather than to either marsupials or placental mammals. Rather high diploid chromosome numbers of about 70 and 63 have been found in the duck-billed platypus *(Ornithorhynchus anatinus)* and the spiny anteater *(Tachyglossus aculeatus)* of the order *Monotremata*. Many small members can be regarded as microchromosomes (MATTHEY, 1949; VAN BRINK, 1959).

It would be of utmost interest to find if the male heterogamety of the XY/XX-type operates in members of the orders *Crocodylia* and *Chelonia* which represent the premammalian lineages. Unfortunately, the heteromorphic sex elements have not been found in these reptiles. No sex-linked gene is known, and the sex reversal experiments have not been performed on any of these species.

f) Extremely High DNA Values Possessed by Certain Amphibians which Suggest the Polyphyletic Origin of Terrestrial Vertebrates

It is known that birds, snakes, and lizards of today are the peripheral branches of one limb which originated from the ancestral reptile *Ornithosuchus;* mammals emerged from the other limb which was started from *Lycaenops*. The fact that surviving members of the class *Reptilia* fall discretely into two groups (one group belonging to the preavian genome lineage and the other group belonging to the premammalian lineage) may be taken as evidence that *Ornithosuchus* and *Lycaenops* of ancient times already belonged to the two different genome lineages.

Reptiles, in turn, were derived from ancient amphibians grouped together as *Labyrinthodonts*. It appears that *Labyrinthodonts* were of many kinds, representing diverse genome lineages. Most, if not all, of the amphibians of today belong to the genome lineages independent from both the preavian and premammalian lineages.

The most comprehensive survey on DNA values of various amphibians was carried out by JOSEPH GALL of Yale University; his results are used here with his kind permission. All the amphibian species surveyed by him showed higher DNA values than that of placental mammals. DNA values demonstrated by tailless amphibians constituting the order *Salientia* were still not as fantastically high as those shown by members of the order *Caudata*.

Within the order *Salientia,* the American toad *(Bufo americanus,* $2n = 22$) representing the suborder *Procoela,* showed a DNA value 137% that of placental mammals. The DNA value of the leopard frog *(Rana pipiens,* $2n = 26$) and the bull frog *(Rana catesbiana,* $2n = 26$) of the suborder *Diplasiocoela* was 200% of the DNA value of placental mammals. In terms of the absolute content, the family *Ranidae* contained 14.6×10^{-9} mg DNA in each diploid nucleus. While these values are high, they show, closely enough, an affinity to the premammalian lineage. It is expected that if a truly extensive survey is done on tailless amphibians of today, a DNA value very similar to that of the premammalian lineage would be found in some of them.

On the contrary, members of the order *Caudata* showed absolutely no affinity to either the premammalian or the preavian lineages. Within this order, the lowest DNA value was found in the newt *(Triturus cristatus,* $2n = 24$) of the suborder *Salamandroidea.* Yet, it was 830% that of placental mammals, and its close relative, *Triturus viridescens* ($2n = 22$) revealed the even higher DNA value of 1300%. The Congo eel *(Amphiuma means,* $2n = 24$) of the same suborder, as well as the mud-puppy *(Necturus maculosus,* $2n = 24$) of the suborder *Proteidae* had the remarkably high DNA value of 2700% that of placental mammals.

Another interesting aspect of tailed amphibian genomes is that two closely related species belonging to the same family often exhibited a remarkable difference in their DNA values. For instance, *Triturus cristatus* and *Triturus viridescens* belong to the same family *Salamandridae,* yet the DNA value of the latter was 50% greater than that of the former, despite the fact that both had very similar diploid complements. JOSEPH GALL found that each lampbrush bivalent of the latter was longer and had more loops than its counter-part of the former. On this basis, he believes that the increase in DNA value is due to regional duplication of chromosomal segments that occurred in *Triturus viridescens.*

It has been shown in the previous chapter that the Z and W, or the X and Y of amphibians are in such a primitive state of evolution that the W or the Y is still a genetical equivalent of the Z or the X. This primitive state of sex chromosomes may permit polyploid evolution to exceptional members of present day amphibians. SAEZ and BRUM (1959) indicated that South American frog species belonging to the family *Ceratophrydae* form a polyploid series. The lowest diploid chromosome number of this series was 22 and the highest, 110. Indeed, MARIA LUISA BEÇAK and her colleagues (1966) found that 44 chromosomes of *Odontophrynus americanus* formed 11 quadri-valents rather than 22 bivalents at meiosis. Thus among amphibians of today, the difference in DNA contents of related species can be attributed to either regional duplication or polyploidization.

Nevertheless, so far as members of the order *Caudata* are concerned, it is clear that they belong to the genome lineage or lineages altogether different from both the preavian and premammalian lineages.

g) Diverse Genome Lineages Found among Fishes

The inevitable conclusion to be drawn from the above survey on DNA values of the four classes of terrestrial vertebrates is that the evolution from Crossopterygian fishes to Labyrinthodont amphibians was polyphyletic. Today, the subclass *Crossopterygii* is represented only by the lung fish of the order *Dipnoi* and the coelocanth of the order *Actinistia.* These surviving members of the lobe-finned fish must represent merely a fraction of the diverse genome lineages which existed in ancient Crossopterygian fishes ancestral to terrestrial vertebrates. Inasmuch as we have no way of obtaining the information on genomes from the fossils, we must turn to members of the ray-finned fish constituting the subclass *Neopterygii* as the source of indirect information on ancient genome lineages.

Our study, although limited to eight species of the class *Pisces,* appeared to confirm the polyphyletic origin of terrestrial vertebrate genomes (OHNO and ATKIN, 1966).

It was found that surviving members of the order *Dipnoi,* the subclass *Crossopterygii,* show close kinship only to tailed amphibians (the order *Caudata).* The DNA value, 3540% that of mammals, was obtained on the South American lung fish *(Lepidosiren paradoxa,* 2n = 38). According to ALFREY et al. (1955), the absolute DNA value for the African lung fish *(Protopterus,* 2n = 34) was 100×10^{-9} mg, which is about 1400% that of mammals. The relatively low diploid chromosome number, the absence of acrocentrics, the enormous size of individual chromosomes, and the very high DNA value found in the lung fish, are all precise characteristics of the genomes maintained by present-day members of the order *Caudata* of the class *Amphibia.*

Fig. 9. The schematic illustration of the diploid chromosome complement of the South American lung fish *(Lepidosiren paradoxa,* 2n = 38), respresenting the order *Dipnoi* of the subclass *Crossopterygii.* Enormity of individual chromosomes can be realized by noting the largest and smallest pairs of human chromosomes drawn in the same scale in a circle. The DNA value is 3540% that of placental mammals

Although the chronology of evolution suggests that the lung fish could not have been the direct ancestor of the tailed amphibians, it is apparent that both belong to the same particular genome lineage. This lineage is not directly related to the main genome lineages which gave rise to tailless amphibians, reptiles, birds, and mammals of today. The diploid chromosome complement of the South American lung fish *(Lepidosiren paradoxa)* is schematically illustrated in Figure 9. The large size of individual chromosomes can be appreciated by noting the attached micron scale.

The DNA values similar to the premammalian and preavian genome lineages were found among members of the subclass *Neopterygii*.

The rainbow trout *(Salmo irideus,* 2n = 58—64) is the anadromous species belonging to the family *Salmonidae* of the order *Isospondyli*. The DNA value, 80% that of mammals, corresponded well with the values possessed by the orders *Crocodylia* and *Chelonia* of the class *Reptilia*. Thus, this species and other members of the family *Salmonidae* may be regarded as belonging to the premammalian lineage. It is not our intention to imply that trouts and mammals constitute one direct line of descent. Our view is that crocodiles, turtles, and mammals of today descend from a particular group of ancient Crossopterygian fish which already possessed the DNA value similar to that of trouts.

In the rainbow trout, extensive postzygotic rearrangements of chromosome arms take place. Although 104 chromosome arms are maintained in each diploid nucleus, different tissues of the same fish yield different karyotypes (OHNO et al., 1965). The metaphase figure shown in Figure 10a is from the liver of a fish and is made of 40 metacentrics and 24 acrocentrics.

The DNA value similar to that possessed by the class *Aves* as a whole, and also by the order *Squamata* of the class *Reptilia,* was found in the goldfish *(Carrasius auratus,* 2n = 96 − 104) of the family *Cyprinidae,* the order *Ostariophysi*. The DNA value obtained in this species was 52% that of mammals. Thus, members of the family *Cyprinidae* may be regarded as belonging to the preavian genome lineage. The diploid chromosome number of 94 was given to this ornamental species by MAKINO (1941). We found that the diploid chromosome number may vary from 96 to 104 within the same individual. It is most likely that postzygotic chromosomal rearrangements similar to those in the rainbow trout occur to a lesser extent in the goldfish. The metaphase figure shown in Figure 10b contains 101 chromosomes, 64 of them metacentrics.

Our study on various members of the subclass *Neopterygii* further revealed the presence of DNA values much smaller than any of the values possessed by terrestrial vertebrates. Our notion that a series of polyploidization of an ancestral vertebrate genome occurred while vertebrates were still in aquatic forms, appeared to be confirmed.

The DNA value of only 30⁰/o that of mammals was obtained on
two members of the order *Perciformes*, the green sunfish (*Lepomis*

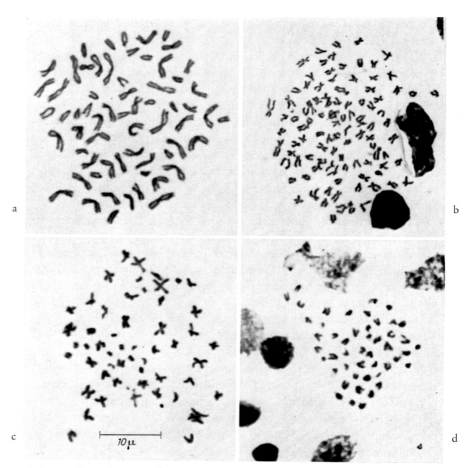

Fig. 10. Somatic metaphase figures of the four diverse species of fishes
representing the four different orders of the subclass *Neopterygii*. a) From
the liver of the rainbow trout (*Salmo irideus*, 2n = 58 – 64) representing the
order *Isospondyli*. The DNA value is 80⁰/o that of placental mammals.
b) The goldfish (*Carrasius auratus*, 2n = 96 – 104) of the order *Ostariophysi*.
The DNA value, 52⁰/o that of placental mammals. c) The discus fish (*Synphysodon aequifasciata*, 2n = 60) of the order *Perciformes*. The DNA value
is 30⁰/o that of placental mammals. d) The hornyhead turbot (*Pleuronichtys
verticalis*, 2n = 48) representing the order *Heterosomata*. The DNA value is
only 20⁰/o that of placental mammals

cyanellus, 2n = 46 – 48) of the family *Centrarchidae* and the discus fish *(Synphysodon aequifasciata,* 2n = 60) of the family *Cichlidae.* The metaphase figure of the discus fish is shown in Figure 10c.

The lowest DNA value, only 20% that of mammals, was found among two diverse groups of fishes. This value was obtained in the swordtail *(Xyphophorus hellerii,* 2n = 48), hornyhead turbot *(Pleuronichthys verticalis,* 2n = 48), and fantail sole *(Xystreurys liolepsis,* 2n = 48).

From the taxonomical point of view as well as from their natural habitats, the swordtail and flatfish are as remotely related as they can be among members of the subclass *Neopterygii.* The swordtail, a Central American fresh-water fish long bred in the aquarium, belongs to the order *Microcyprini,* while two species of the flatfish belong to different families of the order *Heterosomata:* the hornyhead turbot to the right-eyed flounder family, *Pleuronectidae,* and the fantail sole to the left-eyed flounder family, *Bathidae.* Their natural habitat is the ocean bed. The swordtail and the flatfish apparently had identical diploid complements made of 48 acrocentrics gradually declining in size (Fig. 10d), and the lowest DNA value.

We propose to regard these ray-finned fishes as the retainers of the original diploid lineage of ancestral vertebrates. The original diploid lineage then had a DNA value 20% that of mammals. In terms of absolute value, this lineage contained 1.4×10^{-9} mg DNA in each nucleus.

It then follows that the green sunfish and the discus fish belong to the ancient triploid lineage, while the ancient pentaploid lineage is represented by the goldfish, and among terrestrial vertebrates, by lizards, snakes, and birds.

The rainbow trout, crocodiles, and turtles may be regarded as representing the octa- and nonaploid lineages, and placental mammals, the decaploid lineage.

All three constituent polypeptides A, B, and C of the mammalian lactate dehydrogenase have been found to exist in avian species as well as in many of the ray-finned fishes (BLANCO et al., 1964; MARKERT and FAULHAUBER, 1965). These findings on lactate dehydrogenase are in conformity with the view that, in vertebrates, any DNA values above 20% that of placental mammals indicate polyploid lineages. Therefore, sufficient gene duplication has occurred to these genomes. Flatfish of the order *Heterosomata,* on the other hand, revealed the presence of the A-polypeptide only (MARKERT and FAULHAUBER, 1965).

h) Brief Summary of Evolution of Vertebrate Genomes

It appears that gene duplication played a most important role in the evolution of vertebrates. A new gene with a new function arose from a duplicate of the old gene. When the same gene was represented twice within the genome, one redundant gene was allowed to mutate in an independent direction and acquire a new function, while the original function was maintained by the other.

Admittedly, regional duplication of a small number of genes might still be occurring to individual species of higher vertebrates. A series of polyploidization of the ancestral diploid lineage, however, appeared to have occurred while vertebrates were still in aquatic forms nearly 300 million years ago. Among fishes of today, some appear to retain the ancient diploid lineage which contains 1.4×10^{-9} mg DNA per diploid nucleus. Placental mammals, as a whole, appear to belong to the ancient decaploid lineage, while birds represent the ancient pentaploid lineage. Once the chromosomal sex-determining mechanism is well established, no further polyploidization is possible.

As a result, diverse species of placental mammals contain an identical amount of DNA in the diploid complement, 7.0×10^{-9} mg. Speciation within the infraclass *Eutheria* is accomplished almost exclusively by allelic mutations with little change in the total number of gene loci. The same can be said of various avian species. Among reptiles of today, snakes and lizards belong to the preavian pentaploid lineage. Crocodiles and turtles, on the other hand, show close kinship to the decaploid mammalian lineage.

References

ALLFREY, V. G., A. E. MIRSKY, and H. STERN: The chemistry of the cell nucleus. Adv. Enzymol. 16, 411—500 (1955).

ATKIN, N. B., G. MATTINSON, W. BEÇAK, and S. OHNO: The comparative DNA content of 19 species of placental mammals, reptiles, and birds. Chromosoma (Berl.) 17, 1—10 (1965).

— —, and M. C. BAKER: A comparison of the DNA content and chromosome number of 50 human tumours. Brit. J. Cancer 20, 87—101 (1966).

Beçak, W., M. L. Beçak, and H. R. S. Nazareth: Karyotypic studies of two species of South American snakes *(Boa constrictor amarali* and *Bothrops jararaca)*. Cytogenetics 1, 305—313 (1962).

— — —, and S. Ohno: Close karyological kinship between the reptilian suborder *Serpentes* and the class *Aves*. Chromosoma (Berl.) 15, 606—617 (1964).

Beçak, M. L., W. Beçak, and M. N. Rabello: Cytologic evidence of constant tetraploidy in the bisexual South American frog, *Odontophrynus americanus*. Chromosoma 19, 188—193 (1966).

Blanco, A., and W. H. Zinhkam: Lactate dehydrogenases in human testes. Science 139, 601—602 (1962).

— —, and L. Kupchyk: Genetic control and ontogeny of lactate dehydrogenase in pigeon testes. J. exp. Zool. 156, 137—152 (1964).

Fankhauser, G., and R. R. Humphrey: The origin of spontaneous heteroploids in the progeny of diploid, triploid and tetraploid axolotl females. J. exp. Zool. 142, 379—422 (1959).

Galton, M., K. Benirschke, and S. Ohno: Sex chromosomes of the chinchilla; Allocycly and duplication sequence in somatic cells and behavior in meiosis. Chromosoma (Berl.) 16, 668—680 (1965).

Hamerton, J. L., H. P. Klinger, D. E. Mutton, and E. M. Lang: The somatic chromosomes of *Hominoidea*. Cytogenetics 2, 240—263 (1963).

Hollingsworth, M. J.: The metaphase chromosomes of *Crocodilus niloticus*. Cytologia (Tokyo) 22, 412—414 (1957).

Humphrey, R. R., and G. Fankhauser: Structure and functional capacity of the ovaries of higher polyploid (4N, 5N) in the Mexican Axolotl *(Siredon* or *Ambystoma mexicanum)*. J. Morphol. 98, 161—198 (1956).

Ingram, V. M.: The hemoglobin in genetics and evolution. New York: Columbia University Press 1963.

Klinger, H. P.: The somatic chromosomes of some primates *(Tupaia glis, Nycticebus coucang, Tarsius bancanus, Cercocebus aterrimus, Symphalangus syndactylus)*. Cytogenetics 2, 140—151 (1963).

Kobel, H. R.: Heterochromosomen bei *Vipera berus* L. *(Viperidae, Serpentes)*. Experientia 18, 173—174 (1962).

Makino, S.: A karyological study of gold-fish of Japan. Cytologia 12, 96—111 (1941).

Mandel, P., P. Métais et S. Cuny: Les quantités d'acide désoxypentosenucléique par leucocyte chez diverses espéces de Mammiféres. C. R. Acad. Sci. (Paris) 231, 1172—1174 (1950).

Margoliash, E.: Primary structure and evolution of cytochrome c. Proc. Nat. Acad. Sci. 50, 672—679 (1963).

Markert, C. L.: Cellular differentiation—an expression of differential gene function. In: Congenital malformations, pages 163—174. New York: The International Medical Congress 1964.

—, and I. Faulhauber: Lactate dehydrogenase isozyme patterns of fish. J. exp. Zool. 159, 319—332 (1965).

Matthey, R.: Les chromosomes des vertébrés. Lausanne 1949.

MATTHEY, R.: La formule chromosomique et le problème de la détermination sexuelle chez *Ellobius lutescens* Thomas. *Rodentia-Muridae-Microtinae.* Arch. Klaus-Stift. Vererb.-L. **28**, 65—73 (1953).

OHNO, S., J. JAINCHILL, and C. STENIUS: The creeping vole *(Microtus oregoni)* as a gonosomic mosaic. I. The OY/XY constitution of the male. Cytogenetics **2**, 232—239 (1963).

—, C. STENIUS, L. C. CHRISTIAN, W. BEÇAK, and M. L. BEÇAK: Chromosomal uniformity in the avian subclass *Carinatae*. Chromosoma (Berl.) **15**, 280—288 (1964).

— —, E. FAISST, and M. T. ZENZES: Post-zygotic chromosomal rearrangements in rainbow trout *(Salmo irideus* Gibbons). Cytogenetics **4**, 117—129 (1965).

—, and N. B. ATKIN: Comparative DNA values and chromosome complements of eight species of fishes. Chromosoma (Berl.) **18**, 455—466 (1966).

PENNOCK, L. A.: Triploidy in parthenogenetic species of the *Teiid* lizard, Genus *Cnemidophorus*. Science **149**, 539 (1965).

SAEZ, F. A., and N. BRUM: Citogenetica de anfibios anuros de America del Sur. An. Fac. Med. Montevideo **44**, 414—423 (1959).

UZZELL, T. M.: Natural triploidy in salamanders related to *Ambystoma jeffersonianum*. Science **139**, 113—115 (1963).

VAN BRINK, J. M.: L'expression morphologique de la Digamétie chez les Sauropsidés et les Monotrémes. Chromosoma (Berl.) **10**, 1—72 (1959).

Chapter 4

Conservation of the Original X and Homology of the X-linked Genes in Placental Mammals

Comparative DNA values of various vertebrates considered in Chapter 3 have indicated the polyphyletic origin of vertebrate genomes. It appeared that a series of polyploidization of the original vertebrate genome occurred in ancient times. Once the chromosomal sex-determining mechanism was well established, the genome of that lineage was stabilized, and no further gene duplication by polyploidization was possible. Indeed, diverse species of placental mammals gave the identical DNA value of 7.0×10^{-9} mg per diploid nucleus. It then became apparent that in this class extensive speciation from a common ancestor was accomplished almost exclusively by allelic mutation of individual gene loci with little or no change in the total number of gene loci.

Various evidences presented in Chapters 1 and 2, on the other hand, revealed that the differentiation of the X and the Y from a homologous pair was accomplished at the expense of the hetero-gametic sex determiner. While the Y shed Mendelian genes which were originally on it, all these genes were conserved by the X.

Conservation of the original X within the stable genome implies that there should be extensive homology of the X-linked genes among placental mammals, for so-called X-linked genes are the Mendelian genes which were already there when the X was merely a member of an ordinary homologous pair. In this chapter the X-chromosomes of diverse species of placental mammals will be compared.

a) Cytological Evidence of Conservation

Despite the sameness of DNA content, placental mammals of today display an enormous array of karyotypes. MATTHEY (1958) compiled diploid chromosome numbers of 240 species of placental mammals. Although the chromosome number 48 constituted a peak, there was a wide spread at each side of the peak. In the infraclass *Eutheria*, divergence of karyotypes is a feature not only of each order, but also of a certain family. Within the order *Primates*, the diploid number of 80 was found on *Tarsius bancanus* belonging to the suborder *Tarsii* (KLINGER, 1963), while only 34 chromosomes were possessed by two species of spider monkey (*Ateles geoffroyi cucullatus* and *Ateles paniscus chamek*) belonging to the suborder *Simiae* (CHU and BENDER, 1962). Within the rodent family *Muridae*, the diploid number ranged from a high of 78 in *Cricetomys gambianus* of the subfamily *Murinae* (MATTHEY, 1958) to a low of 17 in *Microtus oregoni* and *Ellobius lutescens* of the subfamily *Microtinae* (OHNO et al., 1963; MATTHEY, 1953).

This enormous array of karyotypes reveals the extent to which the original linkage groups of a common ancestor have been shuffled during speciation of placental mammals. While the autosomes have broken and reunited many times, the X-chromosome has not been involved in shuffling, and apparently persisted without substantial change from its beginning. The observation on relative size of the sex chromatin body and the drumsticks of divergent species of mammals gave the first clue to this conservation.

The sex chromatin body is a prominent chromocenter which occurs in the somatic interphase nucleus of the female, but not of the male. A sexual dimorphism is particularly clear in nerve cell nuclei. On the basis of the extensive comparative study made earlier by MOORE and BARR (1953), BARR (1961) has stated, "The sex chromatin of female nuclei is of the same order of size in different representatives of the mammalian class, with mean dimensions of 0.8×1.1 μ."

In the case of granulocytes in blood, a sexual dimorphism is due to the exclusive occurrence in the female of a particular nuclear appendage known as a drumstick. Here, too, DAVIDSON and SMITH (1963) stated that the leukocyte drumsticks of many species of mammals appear to be identical in form and relative size with those in man. Since then, it has been shown that each sex chromatin body represents a single X-chromosome in a positively heteropyknotic state (OHNO et al., 1959). The sameness in size of the sex chromatin body as well as the drumstick is an indication that man ($2n = 46$), the cat ($2n = 38$), the dog ($2n = 78$), cattle ($2n = 60$), the rabbit ($2n = 44$), and many other placental mammals possess the X-chromosome of nearly identical size. The direct measurements of individual X-chromosomes of divergent placental mammals became desirable. In 1964, we devised a method for making two-dimensional measurements of individual metaphase chromosomes (OHNO et al.). Measurements were made in the following way. Photomicrographs of five metaphase figures were selected for each species. Every negative had been made at the same magnification ($2200 \times$). Each negative was placed in the photographic enlarger. The image, magnified to 6300 times, was projected onto a sheet of a particular brand of white typing paper. The outline of each chromosome was carefully traced with a sharp, hard pencil and the images cut out and weighed on a precision balance. The positive identification of individual X-chromosomes at metaphase is possible only in a small number of mammalian species. In most, the best one can do is to determine a particular size group which includes the X. Among those which were subjected to measurements, the X was identified with absolute certainty in the following three species. In the cattle *(Bos taurus,* $2n = 60$) and the dog *(Canis familiaris,* $2n = 78$), the metacentric X stood out from the autosomes, all of which were acrocentrics (AWA et al., 1959; SASAKI and MAKINO, 1962; OHNO et al., 1962). In the donkey *(Equus asinus,* $2n = 62$), the X is the second largest subterminal element in the haploid set (TRUJILLO et al., 1962) (Figs. 11 a, b, and c). Mean absolute sizes of the X of these three species ranged from 4.11 to 4.65 µ², and the X constituted 5.07 to 5.60% of the homogametic (AX) haploid set.

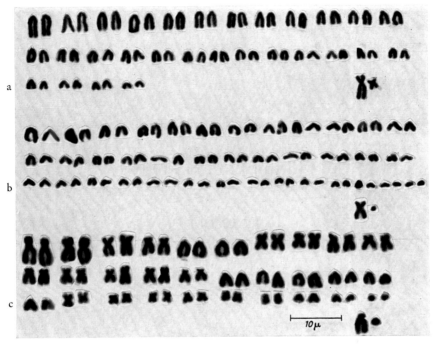

Fig. 11 a—c

Fig. 11. The male diploid chromosome complements of six diverse species of placental mammals, demonstrating the extreme similarity in absolute size of the original-type X. The X and the Y are placed at the extreme right of the bottom row of each karyotype. a) The cattle (*Bos taurus*, 2n = 60) of the order *Artiodactyla*. Both the X and the Y can be identified with absolute certainty. b) The dog (*Canis familiaris*, 2n = 78) representing the order *Carnivora*. The X can be identified. The Y appears to be smaller than the smallest autosome. c) The donkey (*Equus asinus*, 2n = 62) belonging to the order *Perissodactyla*. The X can be identified; the identification of the Y is somewhat tentative. d) Man (*Homo sapiens*, 2n = 46) representing the order *Primates*. The identification of the X is tentative, but the Y can be singled out. e) The cat (*Felis domestica*, 2n = 38) of the order *Carnivora*. Both the X and the Y cannot be identified with absolute certainty. f) The mouse (*Mus musculus*, 2n = 40) belonging to the order *Rodentia*. The identification of the X is uncertain, while the Y can be singled out

Positive identification of the X-chromosome in man, the cat, and the mouse was not possible. In man, the X was thought to be a larger member of the 6—12 group (ROBINSON, 1960). In the cat *(Felis domestica,* 2n = 38), the X was considered to be a smaller member of the 6—11 group (HSU et al., 1963). In the mouse *(Mus musculus,* 2n = 40), with its 40 acrocentric chromosomes declining gradually in size, the X can be the second, the third, fourth, or even the fifth largest element (LEVAN et al., 1962). Reflecting uncertainty of identification, mean absolute sizes of the presumptive X of the above three species ranged from 3.75 to 4.77 μ^2, and the presumptive X comprised 5.10 to 6.45% of the haploid (AX) set (Figs. 11 d, e, and f).

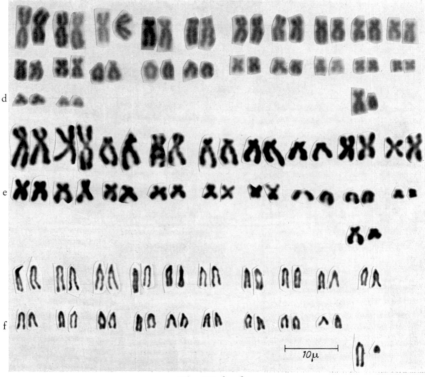

Fig. 11 d—f

Considering the already-proven sameness of DNA content, the above measurements of individual X-chromosomes of man and five other mammalian species clearly furnished the morphological evidence of conservation. In a common ancestor of all the placental mammals, the X must have comprised about 5% of the genome. Despite intense shuffling of autosomal linkage groups which accompanied speciation, diverse species of today apparently preserve the original X of a common ancestor in its entirety.

Within the order *Rodentia*, there is a considerable number of exceptional species which are endowed with X-chromosomes of unusually large sizes. However, the excessive parts of these large X-chromosomes are genetically inactivated so that when only the functional part of the X is considered, it still comprises about 5% of the genome. These exceptional X-chromosomes will be scrutinized later in relation to the dosage compensation mechanism.

b) Problems of Genocopies in Seeking Homology of the X-linked Genes

Conservation of the original X naturally implies extensive homology of the X-linked genes among diverse species of placental mammals. The eleven known X-linked traits of the mouse *(Mus musculus)* are listed in Table 1. The 59 hereditary traits of man for

Table 1. *A catalog of X-borne mutations in the mouse (Mus musculus)*

1. Bent-tail	*(Bn)*	
2. jimpy	*(jp)*	
3. Gyro	*(Gy)*	
4. Tabby	*(Ta)*	
5. Striated	*(Str)*	
6. scurfy	*(sf)*	
7. Mottled	*(Mo)*	
8. Brindled	*(Mo*^{br}*)*	
9. Dappled	*(Mo*^{dp}*)*	} Mottled series
10. Dappled-2	(?)	
11. Tortoiseshell	*(To)*	
12. 26-K	(?)	
13. Blotchy	*(Blo)*	
14. Sex-linked anemia	*(sla)*	
15. Sex-linked histocompatibility antigen	(?)	

which X-linkage is considered proven by McKusick (1962) are compiled in Table 2. Observing Table 1, it may be noted that, except for the X-linked anemia (Falconer and Isaacson, 1962) and the X-linked histocompatibility gene (BAILEY, 1963), all others are morphological traits affecting coat color, hair and skeletal structures.

Table 2. *A catalog of X-borne mutations in man*

1. Partial color blindness, deutan series
2. Partial color blindness, protan series
3. Total color blindness
4. Glucose-6-phosphate dehydrogenase deficiency
5. Xg blood group system
6. Muscular dystrophy, Duchenne type
7. Muscular dystrophy, Becker type
8. Hemophilia A
9. Hemophilia B
10. Agammoglobulinemia
11. Hurler syndrome
12. Late spondylo-epiphyseal dysplasia
13. Aldrich syndrome
14. Hypophosphatemia
15. Hypoparathyroidism
16. Nephrogenic diabetes insipidus
17. Neurohypophyseal diabetes insipidus
18. Low's oculo-cerebro-renal-syndrome
19. Hypochromic anemia (Cooley-Rundles-Falls type)
20. Angiokeratoma diffusum corporis universale
21. Dyskeratosis congenita
22. Dystrophia bullosa hereditaria, typus maculatus
23. Keratosis follicularis spinulosa cum ophiasi
24. Ichthyosis vulgaris
25. Anhidrotic ectodermal dysplasia
26. Amelogenesis imperfecta, hypomaturation type
27. Amelogenesis imperfecta, hypoplastic type
28. Absence of central incisors
29. Congenital deafness
30. Progressive deafness
31. Mental deficiency
32. Börjeson syndrome
33. Spinal ataxia
34. Cerebellar ataxia with extrapyramidal involvement
35. Spastic paraplegia
36. Progressive bulbar paralysis
37. Charcot-Marie-Tooth peroneal muscular atrophy
38. Diffuse cerebral sclerosis (Pelizaeus-Merzbacher)
39. Diffuse cerebral sclerosis (Scholz)
40. Hydrocephalus
41. Parkinsonism

Table 2 (continued). *A catalog of X-borne mutations in man*

42. Ocular albinism
43. External ophthalmoplegia and myopia
44. Microphthalmia
45. Microphthalmia, with digital anomalies
46. Nystagmus
47. Megalocornea
48. Hypoplasia of iris with glaucoma
49. Congenital total cataract
50. Congenital cataract with microcornea
51. Stationary night blindness with myopia
52. Choroideremia
53. Retinitis pigmentosa
54. Macular dystrophy
55. Retinoschisis
56. Pseudoglioma
57. Van den Bosch syndrome
58. Menkes syndrome
59. Albinism-deafness syndrome

Similarly, nearly half of the 59 X-linked traits of man are morphological traits (Table 2). In the case of morphological traits, one cannot place too much confidence in an apparent homology of the two traits, for the two might be genocopies of each other. The two morphological traits of the mouse illustrate the problem of genocopies quite well. The *Tabby (Ta)* is an X-linked mutant gene (FALCONER, 1953), while *crinkled (cr)* is a simple recessive gene located on the linkage group XIV autosome (FALCONER et al., 1951). Yet, the male or the XO-female hemizygous for *Ta* is affected in exactly the same manner as the male or the female homozygous for *cr*. The effects of both genes are manifold, but the most obvious anomalies involve the coat. The coat is thin and has an abnormal texture which gives an affected animal an ungroomed look (Fig. 12). As described in detail by DRY (1926), the coat of the normal mouse consists of long straight guard hairs (about 2%), awls (28%) which are also straight but shorter, and about 70% of zig-zag hairs. In the coat of affected animals, neither guard hairs nor zig-zag hairs are found; it consists entirely of awls which are, however, histologically abnormal. The three main types of hairs on the dorsal part of the normal mouse differ in the number of longitudinal rows of air cells. Guard hairs have two rows, awls may have two or three, while zig-zags have one row in straight parts; but at the constrictions the air space is obliterated and the hair is solid. The awls of affected animals differ from those of normals in that the number of rows of air cells is not constant throughout the length of each hair.

The tail of affected mice is practically naked, and near its tip some kinks are usually seen, reflecting abnormality of skeletal development. The absence of guard hairs and zig-zags in affected adult mice can be traced back to developmental abnormality of hair follicle formation. During fetal development of the affected, no follicles are formed between the 14th and 17th days. Although the formation of hair follicles starts on the 18th day and continues until the time of birth, no further follicles are laid following birth. The coat of affected mice consists of awls only because these are the hairs formed between the 18th day and parturition, the only period of follicle formation in these mice. Furthermore, GRÜNEBERG (1965) has recently reported that teeth of the *Ta*-hemizygote and the *cr*-homozygote are again affected in the same manner.

It is fortunate that both *Ta* and *cr* are found in the same species. If the autosomally linked *crinkled* were known in the mouse, and the X-linked *Tabby* found in another rodent species, the two traits could be used as an argument which would speak against the conservation of the original X.

Fig. 12. An appearance of a male mouse hemizygous for the X-linked mutant gene *Tabby* (left) is compared with that of a wild-type female CBA mouse (right). The phenotype of a hemizygous *Tabby* mouse is indistinguishable from that of a male or female mouse homozygous for the autosomal recessive gene *crinkled*

Once it is realized that a gene is a particular DNA molecule which specifies the amino acid sequence of a polypeptide it produces, the phenomenon of genocopies can be readily understood. Let us assume that two enzymes, *a* and *b*, which catalyze successive steps of a metabolic pathway are produced by two independent genes on different chromosomes. The enzyme *a* converts a metabolite *A* to *B*, then the enzyme *b* acts and further converts *B* to *C*. A morphological trait which is due to deficiency of the metabolite *C* can be a result of a defective mutation at the gene locus for the enzyme *b*, but a defective mutation at the gene locus for the enzyme *a* also results in deficiency of the metabolite *C*. Thus, an X-linked trait of man or the mouse, which is known as a syndrome (a set of symptoms), should be dealt with extreme caution in seeking homology with a similar trait of another species.

c) The Problem of Atavism

Another difficulty which arises in seeking homology of X-linked traits between diverse species of mammals is the problem of atavism. A good example can be found in the colorblindness traits of man. In man, partial colorblindness (dichromatism) of both protanopia and deuteranopia series (FRANCESCHETTI and KLEIN, 1957) as well as total colorblindness (achromatism) of a certain type (OPITZ, 1961), have been shown to be X-linked.

While many lower vertebrates and birds are equipped with excellent trichromatic color vision (WARNER, 1931; WOJTSUSIAK, 1933; HAMILTON and COLEMAN, 1933), numerous experiments on color vision of mammals other than higher primates have indicated that, in this entire group of animals, color vision is very rudimentary at best (PARSONS, 1924). Protoinsectivores, a common ancestor of placental mammals, probably had nocturnal habits, and this must be the reason for the widespread occurrence of achromatic (black and white) vision among mammals of today.

Several color vision theories have offered outlines of the probable evolutionary development of human trichromatic vision. LADD-FRANKLIN (1929) proposed three evolutional stages, namely, achromatic vision, blue-yellow vision, and finally blue-green-red vision.

The comparative studies made on color vision of various members of the order *Primates* appear to confirm the LADD-FRANKLIN theory in principle. Lower species belonging to the suborder *Lemuroidea* apparently maintain achromatic vision (BIERENS DE HAAN and FRIMA, 1930), while GRETHER (1939) found that the ringtail monkey

(Cebus capucinus and *Cebus unicolor)* of the family *Cebidae* representing the New World *(Platyrrhina)* monkey possessed dichromatic vision of the protanopic type. Wave-length discrimination of the ringtails was significantly poorer in the red and yellow region, but was about equal to man in the blue-green region. On the other hand, the Old World *(Catarrhina)* monkeys, such as baboons and rhesus, exhibited trichromatic vision. Wave-length discrimination did not differ significantly from that of man.

In the evolution of primates, the gene for achromatic vision apparently represented the original wild-type allele, while the gene for protanopic dichromatic vision arose as a mutant, as did the gene for trichromatic vision. Thus, partial and total colorblindness of man are to be regarded as a recurrence of atavistic traits. In the case of these hereditary traits of man, homology should be sought by searching for mutant individuals with dichromatic or trichromatic vision among mammalian species other than higher primates. If the X-linkage of the trait for chromatic vision is established in other mammalian species, the homology with the colorblindness traits of man can be considered proven.

d) Problems in Discriminating Structural Defects from Regulatory Defects

In the case of hemophilia A and B of man, a particular protein which is deficient in each trait has been identified. Thus, we are very close to identifying the direct products of the two gene loci on the X. Hemophilia A is the result of a hereditary deficiency in antihemophilic globulin (AHG or Factor VIII), while hemophilia B or Christmas disease is the result of a hereditary deficiency in plasma thromboplastic component (PTC or Factor IX).

However, when the presence of a particular gene on the X is implied only through its apparently defective mutant allele, there is an inherent danger of mistaking a regulatory gene locus for a structural gene locus. An apparent absence of a functional protein may be due to active synthesis by a mutant gene of a functionless protein molecule. If this is the cause of hemophilia A and B, the structural genes for AHG and PTC are definitely located on the human X. Conversely, an apparent absence of a functional protein may be due to repression of the normal structural gene by an abnormal regulator gene. Although the structural gene for AHG may reside on the X, the regulatory gene which controls the production of AHG may be on an autosome. In the bacterium *(Escherichia coli)* in which the concept of the genetic regulatory mechanism has been developed (JACOB and MONOD, 1961), it has been shown that the regulatory gene need not be closely linked to the structural gene it controls. Von Willebrand's disease (vascular hemophilia) is an autosomally linked hereditary trait of man, yet deficiency of AHG occurs in this trait as well (GRAHAM, 1959). If an autosomally linked AHG deficiency had been found in other mammals, it might have been used as an argument against the conservation of the primitive X.

e) Confusion which can Arise from the Polymeric Nature of many Enzymatic and Non-enzymatic Proteins

Another danger in using defective hereditary traits for the study of homology lies in the fact that many of the enzymatic and non-enzymatic protein molecules are made of two different component polypeptides produced by two unlinked genes. Lactate dehydrogenase, hemoglobin, and γ-globulin are such examples, as stated previously. It may be that human AHG is made of two different polypeptides, one produced by the X-linked structural gene and the other by an autosomally linked structural gene. The defective mutation of either of the two genes would result in deficiency of AHG. Hemophilia A may be due to a defective mutation of the X-linked structural gene, and von Willebrand's disease to that of an autosomal gene.

f) Problems Concerning Cell Differentiation Process

Unless careful examination of various symptoms which characterize a particular defective trait is made, structural defects can also be confused with defects which affect the cell differentiation process. It has been shown in man that the autosomal locus for *Gm* allotypes marks the site of structural genes for H-chain of γ-globulin molecules, while the other autosomally linked gene locus for *Inv* allotypes indicates the site of structural genes for L-chain (STEINBURG, 1962). Clearly, the X-linked agammaglobulinemia of man is not a structural defect. Deficiency of γ-globulin is a consequence of a failure of lymphoid cells to differentiate into plasma cells. Without the formation of plasma cells, there can be no γ-globulin production, even if lymphoid cells contain the normal structural genes for both H- and L-chains.

g) X-linkage of Glucose-6-phosphate Dehydrogenase Gene in Man and Other Mammals

From the above, it becomes clear that most of the known X-linked traits of the mouse *(Mus musculus)* and man listed in Tables 1 and 2 are not very useful in seeking the exact homolgy of X-linked genes among diverse species of placental mammals. What is needed is a structural gene whose direct product can be recognized and measured. The structural gene for an enzyme, glucose-6-phosphate dehydrogenase, is such an ideal gene. The X-linkage of this gene has been shown in man, the horse, the donkey, and two species of wild-hares of Europe. Glucose-6-phosphate dehydrogenase (G-6-PD) catalyzes the first step of pentose phosphate shunt of carbohydrate metabolism by converting glucose-6-phosphate to 6-phosphogluconic acid in the presence of triphosphopyridine nucleotide (TPN). Reduced triphosphopyridine nucleotide (TPNH) is generated by this reaction. G-6-PD is a trimer having the molecular weight of about 120,000. The full enzymatic activity requires a trimer to be bound with TPN (KIRKMAN and HENDRICKSON, 1962; MOTULSKY and YOSHIDA, personal communication).

In man, it was first found that various hereditary deficiencies of this enzyme are inherited as X-linked traits. It appears that there are at least three different defective mutations involving G-6-PD: (a) G-6-PD deficiency in Africans which is clinically recognized as a sensitivity to an antimalarial drug, "primaquine"; (b) favism in Mediterranean peoples (G-6-PD deficiency manifests itself as a sensitivity to ingestion of a certain type of beans); and (c) non-spherocytic hemolytic anemia due to G-6-PD deficiency (CHILDS et al., 1958; BEUTLER, 1960; MARKS and GROSS, 1959; KIRKMAN et al., 1960).

When only the defective mutations were known, it remained uncertain whether or not the human X in fact carried the structural gene for G-6-PD. It was fortunate that a few electrophoretic variants of G-6-PD were subsequently found (BOYER et al., 1962; KIRKMAN and HENDRICKSON, 1963).

A change in the base sequence of a DNA molecule, which is the structural gene, can result in the amino acid substitution of the polypeptide chain it produces. If this occurs, a single neutral (monoamino, monocarboxyl) amino acid may be replaced by an acidic (monoamino, dicarboxyl) amino acid. As a result, a polypeptide produced by a mutant gene gains one more negative charge. Each G-6-PD molecule, if it is a trimer, gives to a mutant G-6-PD three more negative charges per molecule. A mutant G-6-PD can readily be distinguished from the wild-type G-6-PD by electrophoresis. If an electrophoresis is run at a pH more alkaline than the isoelectric points of both, both enzyme molecules would move toward the positive pole (anode), but a mutant molecule which has more negative charges than the wild-type molecule would move further toward the anodal direction. If, on the other hand, an electrophoresis is run at a pH which is more alkaline than the isoelectric point of a mutant G-6-PD, but more acidic than that of the wild-type G-6-PD, a mutant G-6-PD molecule would still move toward the anodal direction, but the wild-type molecule would now move toward the negative pole (cathode).

If such an amino acid substitution occurred in a non-critical part of the polypeptide chain, it would not hamper the enzymatic efficiency of a mutant molecule. Thus, a species may come to maintain several such alleles at the structural gene locus for G-6-PD. The products of these alleles maintain normal enzymatic activity, but display different isoelectric points. These products are known as the electrophoretic variants of G-6-PD.

In man, the wild-type or species specific G-6-PD is known as the electrophoretic variant B. In addition, the faster moving (more negatively charged) variant A occurs with high frequency among Africans. Although hereditary G-6-PD deficiency of Africans is due to a defective mutation probably superimposed on the variant A allele (A^-), the variant A itself is enzymatically as efficient as the variant B.

When these two electrophoretic variants became known in man, the X-linkage of the structural gene for G-6-PD had been established beyond doubt. The heterozygous state *(A/B)* existed only in the normal XX-female and the abnormal XXY-male, but never in the normal XY-male and the abnormal XO-female. The mating between a phenotype *(A)* male and a phenotype *(B)* female invariably gave the heterozygous *(A/B)* state to all of their daughters, but all the sons who received a single X from the mother were phenotype *(B)* (Fig. 13a).

The presence of electrophoretic variants of G-6-PD in man encouraged us to make a systematic test for the X-linkage of this enzyme on diverse species of placental mammals using starch gel electrophoresis as a tool. Initial endeavors were not very rewarding. Not only many breeds of dogs tested demonstrated the one identical type of G-6-PD, the G-6-PD of other members of the family *Canidae,* such as the wolf, the coyote, the jackal, and the dingo (Australian wild dog) were indistinguishable from that of the domestic dog as well. The same disappointing results were also obtained on various breeds of the horse as well as on various inbred strains of the mouse.

Faced with an apparent absence of intraspecific allelic polymorphism at the G-6-PD structural gene locus in these mammals, we decided to test the X-linkage of this enzyme through reciprocal interspecific crosses. The mule is a cross between the male donkey *(Equus asinus)* and the female horse *(Equus caballus).* The reciprocal cross of the above is known as the hinny. It was found that the species specific G-6-PD of the donkey can readily be distinguished from the species specific G-6-PD of the horse. Because the former molecule is considerably more negatively charged than the latter, the former moved much farther toward the anodal direction than the latter when electrophoresis was performed at pH 8.6 as well as at pH 7.0. When reciprocal hybrids between the two were tested, all the female mules as well as female hinnies demonstrated the coexistence of the two parental types of G-6-PD (Fig. 13b), but all the male mules possessed the maternally derived horse G-6-PD only. The male hinny, on the contrary, demonstrated only the donkey G-6-PD and not a trace of the horse G-6-PD. Thus, the X-linkage of the structural gene for G-6-PD has been proven on the horse as well as the donkey (TRUJILLO et al., 1965; MATHAI et al., 1966).

There are two species of wild hares in Europe, the common wild hare *(Lepus europaeus)* and the variegating hare of the north *(Lepus timidus).* The reciprocal crosses between the two species have been successfully obtained by INGEMAR GUSTAVSSON of the Royal Veterinary College, Sweden. First, when electrophoresis was performed at pH 7.0, it was established that the species specific G-6-PD of *Lepus europaeus* was considerably faster moving toward the anodal direction than that of *Lepus timidus.* Again, the female hybrids invariably demonstrated the coexistence of both parental types of G-6-PD, but each male hybrid displayed only the maternally derived G-6-PD (Fig. 13c). The X-linkage of the G-6-PD structural gene was also established in members of the order *Lagomorpha* (OHNO et al., 1965).

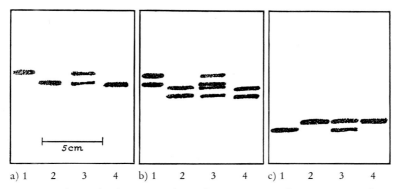

a) 1   2   3   4   b) 1   2   3   4   c) 1   2   3   4

Fig. 13. A schematic demonstration of zymograms of an enzyme glucose-6-phosphate dehydrogenase (G-6-PD) as revealed by starch gel electro-phoresis. These studies revealed the X-linkage of the structural gene for erythrocyte G-6-PD in man (Fig. 13 a), the horse and the donkey (Fig. 13 b) and two species of wild hares of Europe (Fig. 13 c). a) From left to right, 1) a father hemizygous for a variant enzyme (A^+); 2) a mother homozygous for the wild-type (B^+); 3) a daughter heterozygous (A^+/B^+) and 4) a son hemizygous (B^+). b) From left to right, 1) a father donkey hemizygous for the wild-type (D); 2) a mother horse homozygous for the wild-type (H); 3) a daughter mule heterozygous (D/H) and 4) a son mule hemizygous (H). c) From left to right, 1) a father variegating hare *(Lepus timidus)* hemi-zygous for the wild-type (T); 2) a mother common hare *(Lepus europaeus)* homozygous for the wild-type (E); 3) a daughter hybrid heterozygous (T/E) and 4) a son hybrid hemizygous for (E)

Man belongs to the order *Primates,* while the horse and the donkey belong to the order *Perissodactyla* and the hares to the order *Lagomorpha.* These three orders are as remotely related as they can be among placental mammals. The establishment of the X-linkage of the G-6-PD structural gene on these five diverse species speaks well of the notion that the original X of a common ancestor has been preserved in its entirety by diverse species of placental mammals.

The X-linkage of the G-6-PD structural gene has also been established in the fruit fly *(Drosophila)* (YOUNG et al., 1964). This, however, appears to be a pure coincidence. The X-chromosome of *Drosophila* comprises nearly 30% of the genome; thus, any gene of this insect has a one-in-three chance of being X-linked. Indeed, the structural gene for 6-phosphogluconate dehydrogenase (6-PGD) of *Drosophila* has also been shown to be on the X (KAZAZIAN et al., 1965). In man (FILBES and PARR, 1963) and in the deer mouse *(Peromyscus maniculatus)* (SHAW, 1965), 6-PGD which catalyzes the next step after G-6-PD of pentose phosphate shunt, has been shown to be autosomally inherited.

As the corollary to assumed homology of X-linked genes among diverse species of placental mammals, it is expected that what is autosomally inherited in one species should not be X-linked in another.

SHAW and BARTO (1965) have recently shown that in liver and other tissues, but not in erythrocytes of the deer mouse *(Peromyscus maniculatus)*, in addition to a major component of G-6-PD which was presumably X-linked, there was a small amount of the slower moving component of G-6-PD which was autosomally inherited. We have recently shown that this autosomal G-6-PD of the deer mouse has a different property from the X-linked enzyme in that it performs both as G-6-PD and as Galactose-6-PD with equal efficiency (Fig. 14b). It was further found that a similar component with both G-6-PD and Gal-6-PD activities also occurs in the liver of man, cattle, and other placental mammals.

When vertical starch gel electrophoresis was continued for 16 hours at $4°$ C with a gradient of 4 volt/cm, using pH 8.6 borate buffer, most human liver samples gave a single narrow band of this enzyme at the position 13—15 mm from the starting point. The X-linked variant (B) G-6-PD, on the other hand, migrated 70 to 75 mm toward the anodal direction. These persons were obviously homozygous for the wild-type species specific allele at the gene locus for this Gal-6-PD component (Fig. 14a). However, there was one male in our series who demonstrated three evenly spaced bands of this component, the slowest moving of the three corresponding to a single band of Gal-6-PD demonstrated by the others. It may be that this male was heterozygous for a mutant allele which produces a faster moving variant of the Gal-6-PD component. The middle band can be interpreted to represent a hybrid dimer enzyme. If this interpretation is correct, the autosomal linkage of this component with both G-6-PD and Gal-6-PD activity is implied in man as well (OHNO et al., 1966; SHAW, 1966). This autosomally inherited G-6-PD of the deer mouse has been shown to form a hybrid dimer enzyme in hetero-zygotes (SHAW and BARTO, 1965).

The X-linked G-6-PD and the autosomally inherited hexose-6-phosphate dehydrogenase differ in other aspects as well. The former is found in supernatant fraction, while the latter is found in micro-somal fraction of cells. The former molecule is a trimer according to MOTULSKY and YOSHIDA, while the latter molecule is apparently a dimer.

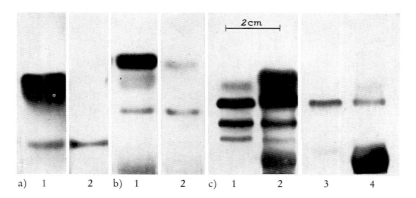

Fig. 14. Photographs of starch gel plates demonstrating the presence of a new autosomally inherited hexose-6-phosphate dehydrogenase with both G-6-PD and Gal-6-PD activities in liver extracts of man (Fig. 14 a), the deer mouse *(Peromyscus maniculatus)* (Fig. 14 b) and the rainbow trout *(Salmo irideus)* (Fig. 14 c). a) 1) is stained for G-6-PD. In addition to the intensely stained X-linked G-6-PD, a narrow slower moving band is noted. 2) is stained for Gal-6-PD. Only a slower moving band shows activity. b) 1) is stained for G-6-PD and 2) is for Gal-6-PD. The autosomal in-heritance of a narrow slower moving band has been proven. c) The rainbow trout contains four isozymes in erythrocytes and five isozymes in liver. 1) and 3) are from erythrocyte extracts, while 2) and 4) are from liver extract. 1) and 2) are stained for G-6-PD and 3) and 4) for Gal-6-PD. The slowest moving narrow band which is present in liver but not in erythro-cytes represents a new hexose-6-phosphate dehydrogenase with both G-6-PD and Gal-6-PD activities

Our study on the rainbow trout *(Salmo irideus)* indicated that this microsomal hexose-6-PD arose very early in the phylogeny of vertebrates. As shown in Fig. 14c, four bands of G-6-PD were distinguished in extracts from erythrocytes of the rainbow trout. The trout apparently has two independent gene loci coding for subunits of G-6-PD of supernatant fraction. Each G-6-PD molecule being a trimer, the products of two gene loci, A and B, are expected to form four isozymes, A_3, A_2B, AB_2 and B_3, in the $1:3:3:1$ ratio. In liver, an additional band which was slower moving than any of the four was found. While no Gal-6-PD activity was detectable in erythrocyte extracts, the slowest moving band from liver showed very intense Gal-6-PD activity. Thus, there is no doubt that the autosomally inherited microsomal hexose-6-PD of placental mammals has a very early evolutional origin.

It is of interest to note that while placental mammals apparently have only one X-linked gene locus for G-6-PD of supernatant fraction, the rainbow trout and other salmonoid fish have two gene loci for this enzyme. As stated in the previous chapter, the DNA value of salmonoid fish is relatively high, being 80% that of placental mammals. Salmonoid fish can be considered as tetraploid in relation to clupeoid fish such as anchovies and herrings. The former has 100 to 104 chromosome arms in the diploid complement, while the latter has 48 chromosome arms and the DNA value 40% that of placental mammals. Wild-type clupeoid fish demonstrate only a single band of G-6-PD of supernatant fraction. It is most likely that two separate gene loci of salmonoid fish correspond to two alleles at one gene locus of clupeoid fish. The above is a good example of gene duplication by polyploid evolution.

The study on the structural gene for glucose-6-phosphate dehydrogenase was indeed very rewarding. It not only supported the notion that what is X-linked in one species should also be X-linked in other species of placental mammals, but also advocated the presence of an autosomally inherited microsomal hexose-6-PD in diverse species of placental mammals. Furthermore, the study on fish suggested that G-6-PD of supernatant and microsomal hexose-6-PD arose independently of each other.

h) X-linkage of Hemophilia A and B in Man, the Dog, and Possibly in the Horse

As stated previously, human hemophilia A is a hereditary deficiency in antihemophilic globulin (AHG or Factor VIII), while hemophilia B or Christmas disease is the result of a hereditary deficiency in plasma thromboplastic component (PTC or Factor IX). The X-linkage of both defects has been established.

The fact that the autosomally inherited defect known as von Willebrand's disease (vascular hemophilia) also demonstrates a deficiency of AHG raised some apprehension about the possibility that a regulatory gene defect of AHG production in other species may be mistakenly regarded as homologous to human hemophilia A which appears to be a structural gene defect.

Nevertheless, it has been shown that hemophilia is also X-linked in the dog, where two varieties of X-linked hemophilia, A and B, have been found. In each type, the physiologic defect is seemingly identical in man and the dog. Furthermore, a pedigree exists which suggests that hemophilia A might also be X-linked in the horse.

Since 1936 it has been known that, in the dog, a clotting defect which closely resembles hemophilia A (classic hemophilia) of man is also inherited as an X-linked recessive trait. This defect has been described in the aberdeen terrier, greyhound, and a few other breeds. In each instance, when an apparently healthy bitch was bred to a normal dog, half of the male puppies developed this defect, while all the female puppies were apparently normal. This defect, however, was transmitted to the next generation through half of these apparently healthy female puppies. HUTT and his colleagues (1948) have accumulated hematological evidences which suggest that this clotting defect is indeed due to deficiency of AHG. This suggestion was further substantiated by the more detailed work of GRAHAM et al. (1949).

Another X-linked clotting defect which is apparently identical with human hemophilia B (Christmas disease) has been found in a family of line, bred cairn terriers. This defect was found to be in the formation of plasma thromboplastin which was corrected by the addition of human serum from hemophilia A, but human hemophilia B serum had no effect. Thus, it is well established that this canine X-linked defect is indeed due to deficiency of PTC (MUSTARD et al., 1960). The clotting defect apparently identical with human hemophilia A has also been found in a male thoroughbred foal. The consistently low level of AHG activity in the foal's blood was raised by human AHG infusion with good clinical response. Furthermore, the breeding record of the dam of the propositus strongly suggested the X-linkage of hemophilia A in the horse. This apparently healthy dam had been bred to eight different stallions. There were two abortions, and of four other male foals born to her, two died in early life, though no information could be obtained on the cause of death. All three female foals, on the other hand, survived normally (NOSSEL et al., 1962). It is of utmost interest to know if the defect was transmitted to male foals of the next generation through any of these three female foals.

As stated previously, an autosomally inherited trait known as von Willebrand's disease (vascular hemophilia) of man also demonstrates an apparent deficiency of AHG. Hemophilia of the pig which is inherited as an autosomal recessive trait is probably homologous to von Willebrand's disease (MUHRER et al., 1942).

Several observations (BIGGS and MATHEWS, 1963) are pertinent to the nature of the AHG defect in von Willebrand's disease. Blood from a patient with hemophilia A will correct the clotting defect in von Willebrand's disease, although the converse is not true. Blood from a patient with von Willebrand's disease will not correct the clotting defect in hemophilia A. Whether or not the swine hemophilia is homologous to von Willebrand's disease of man can be tested. If human hemophilia A blood corrects the clotting defect of the swine hemophilia, the homology is established. No confusion should be made between the autosomally linked hemophilia of the pig and the X-linked hemophilia A of man, the dog, and quite possibly the horse.

i) X-linkage of Anhidrotic Ectodermal Dysplasia in Man and Cattle

Anhidrotic ectodermal dysplasia is a syndrome. CHARLES DARWIN apparently had the knowledge of the occurrence of this defect among men of Sind in Pakistan; its X-linkage has since been well established. An affected male is usually toothless and hairless. Because the sweat glands are non-functional, he is unable to sweat.

DRIEUX et al. (1950) described an apparently identical X-linked trait in cattle as "Hypotrichose congénitale avec anodontie, acérie et macroglossie". Hemizygous males were born toothless and hairless, with cystic sweat glands lacking secretory tubules.

When an apparently healthy cow of Maine-Anjou-Normandy breed was bred to a bull of Charollais breed, she produced two male calves affected with this syndrome and one apparently healthy female calf. This defect was obviously not transmitted by the Charollais bull as his 180-odd calves by other cows were all unaffected. Upon reaching maturity, an apparently healthy female who was a full sib of the two affected males was bred to another bull. One of the three male calves she produced was again affected with this syndrome. Thus, the X-linkage of this defect in the cattle can be considered well proven.

j) Other Examples of Possible Homology between X-linked Genes of Diverse Species of Mammals

A few more examples of possible homology which are described below should be considered merely as suggestive evidence and nothing more.

X-linked anemia in man was first described by COOLEY, who also first described thalassemia. In a family he studied, 19 males in five generations were affected with transmission through unaffected females (COOLEY, 1945). X-linked anemia of man belongs to the category of sideroachrestic or iron-loading anemia. Of course, there are many causes for this category of anemia. The defect is essentially due to the erythropoietic system's inability to utilize iron. The features include elevation of serum iron level, abundance of siderocytes in peripheral blood after splenectomy, and hemochromatosis (BYRD and COOPER, 1961).

X-linked anemia of the mouse *(Mus musculus)* reported by FALCONER and ISAACSON (1962) may be homologous to the above defect in man. Affected hemizygous males display mild hypochromic anemia which is not due to iron deficiency. Anemia is most pronounced during neonatal life (GREWAL, 1962). In our experience, serum iron level appears to be significantly elevated, but a significant number of siderocytes are not seen in peripheral blood after splenectomy, and hemochromatosis is not observed. The absence of hemochromatosis, however, may merely be due to the extremely short normal life span of the mouse compared to man.

As mentioned earlier in this chapter, the X-linked mutant gene *Tabby* of the mouse *(Mus musculus)* inhibits the formation of two of the three types of hair follicles normally formed by the mouse. Thus, the coat of an affected hemizygous male is made entirely of awls; neither guard hairs nor zig-zag hairs are present. The hetero-zygous female invariably demonstrates the mosaic phenotype. Vertical stripes of affected *Tabby* areas are seen in the background of wild-type fur (FALCONER, 1953).

The X-linked trait in Holstein-Friesian cattle described as "streaked-hairlessness" may be homologous to *Tabby*. This defect in the hemizygous state is apparently lethal, but it can be surmised that affected hemizygous males, if they are born, should be completely hairless. The cows which are heterozygous for this trait demonstrate hairless vertical streaks along the entire length of the body. These streaks are especially noticeable in the dorsal region. The most severely affected heterozygous cow is nearly devoid of hair over the thurls (ELDRIDGE and ATKESON, 1953).

If the normal coat of the cattle constitutionally lacks the type of hair follicles which correspond to the awl hair follicles of the mouse, a mutation homologous to *Tabby* of the mouse should give hairless vertical streaks to the female cattle which is heterozygous for this mutation.

It cannot, however, be readily explained by the assumption of the homology between the two why the hemizygous state for *Tabby* of the mouse is viable, while the hemizygous state for "streaked hairlessness" of cattle is lethal.

In the mouse *(Mus musculus)*, the presence of the X-linked gene for histocompatibility antigen has recently been demonstrated (BAILEY, 1963). When the cross was made between two certain inbred strains of the mouse, the F_1 male rejected the skin donated by his sister. This rejection can only be due to the reaction of the F_1 male against the histocompatibility antigen produced by the paternally derived X of the F_1 female. All the autosomally inherited histocompatibility antigens of the F_1 female are not foreign to her brother.

The histocompatibility antigens are, by definition, distributed on the cell surface, and it is known that autosomally inherited histocom-patibility antigens of the mouse are present on the cell surface of erythrocytes. Thus, they can be regarded as blood group genes.

It is possible that the X-linked (X_g) blood group system of man (MANN et al., 1962) and the X-linked histocompatibility system of the mouse are homologous.

k) Brief Summary on Conservation of the Original X and Universality of the X-linked Genes among Placental Mammals

The evidence presented in this chapter tends to support the notion that the original X-chromosome of a common ancestor has been preserved in its entirety by diverse species of placental mammals of today. Despite the enormous array of karyotypes displayed by various species of mammals, in the great majority, including man and the mouse *(Mus musculus)*, the X is nearly identical in size, comprising little more than 5% of the genome.

Obviously, some of the original genes on the primitive X of a common ancestor mutated in different directions during the extensive speciation that followed, acquiring such varied functions along the way that their common heritage is no longer apparent. In other instances, the products of corresponding genes in different species continue to perform the same function and can still be recognized as the same gene. Indeed, the structural gene for a major component of an enzyme, glucose-6-phosphate dehydrogenase, is situated on the X-chromosome not only of man, but also of the horse, the donkey, and wild hares of Europe. Similarly, from hemophilia A and B, it can be implied that the structural gene for a component polypeptide of antihemophilic globulin (AHG or Factor VIII) as well as the structural gene for plasma thromboplastin component (PTC or Factor IX) are both X-linked not only in man, but in the dog as well, and possibly in the horse. While there are more examples which suggest homology between X-linked genes of diverse species, there is not an evidence which shows that what is X-linked in one species is autosomally inherited in others.

References

Awa, A., M. Sasaki, and S. Takayama: An *in vitro* study of the somatic chromosomes in several mammals. Jap. J. Zool. **12**, 257—265 (1959).

Bailey, D. W.: Mosaic histocompatibility of skin grafts from female mice. Science **141**, 631—634 (1963).

Barr, M. L.: Das Geschlechtschromatin (The Sex Chromatin). In: Die Intersexualität, pp. 50—73. Ed. by Claus Overzier. Stuttgart (Germany): Georg Thieme Verlag 1961.

Beutler, E.: Drug-induced hemolytic anemia (primaquine sensitivity). In: The metabolic basis of inherited disease, pp. 1031—1067. Eds. J. B. Stanbury, J. B. Wyngaarden, and D. S. Fredrickson. New York: MacGraw-Hill Book Company Inc. 1960.

BIERENS DE HAAN, J. A., and M. J. Frima: Versuche über den Farbensinn der Lemuren. A. vergl. Physiol. **12**, 603—631 (1930).

BIGGS, R., and J. M. MATHEWS: The treatment of haemorrhage in von Willebrand's disease and the blood level of Factor VIII (AHG). Brit. J. Haematol. **9**, 203—214 (1963).

BOYER, S. H., I. H. PORTER, and R. WEILBOECHER: Electrophoretic heterogeniety of glucose-6-phosphate dehydrogenase and its relationship to enzyme deficiency in man. Proc. Natl. Acad. Sci. U.S. **48**, 1868—1876 (1962).

BYRD, R. B., and T. COOPER: Hereditary iron-loading anemia with secondary hemochromatosis. Ann. intern. Med. **55**, 103—123 (1961).

CHILDS, B., W. H. ZINKHAM, E. A. BROWNE, E. L. KIMBRO, and J. V. TORBERT: A genetic study of a defect in glutathione metabolism of the erythrocyte. Bull. Johns Hopkins Hosp. **102**, 21—37 (1958).

CHU, E. H. Y., and M. A. BENDER: Cytogenetics and evolution of primates. Ann. N. Y. Acad. Sci. **102**, 253—266 (1962).

COOLEY, T. B.: Severe type of hereditary anemia with elliptocytosis: interesting consequence of splenectomy. Amer. J. med. Sci. **209**, 561—568 (1945).

DAVIDSON, W. M., and D. R. SMITH: The nuclear sex of leukocytes. In: Intersexuality. Edited by C. Overzier. London: Academic Press Ltd. 1963.

DRIEUX, H., M. PRIOUZEAU, G. THIERY et M. L. PRIOUZEAU: Hypotrichose congénitale avec anodontie, acérie et macroglossie chez la veau. Rec. Méd. vét. **126**, 385—399 (1950).

DRY, F. W.: The coat of the mouse. J. Genet. **16**, 287—340 (1926).

ELDRIDGE, F., and F. W. ATKESON: Streaked hairlessness in Holstein-Friesian cattle: A sex-linked lethal character. J. Hered. **44**, 265—271 (1953).

FALCONER, D. S., D. S. FRASER, and J. W. B. KING: The genetics and development of "crinkled", a new mutant in the house mouse. J. Genet. **50**, 324—344 (1951).

— Total sex-linkage in the house mouse. Z. indukt. Abstamm.- u. Vererb.-L. **85**, 210—219 (1953).

—, and J. H. ISAACSON: The genetics of sex-linked anaemia in the mouse. Genet. Res. **3**, 248—250 (1962).

FILBES, R. R., and C. W. PARR: Human red blood cell phosphogluconate dehydrogenases. Nature **200**, 890 (1963).

FRANCESCHETTI, A., and D. KLEIN: Two families with parents of different types of red-green blindness. Acta genet. (Basel) **7**, 255—259 (1957).

GRAHAM, J. B., J. A. BUCKWALTER, L. J. HARTLEY, and K. M. BRINKHAUS: Canine hemophilia: observations on the course, the clotting anomaly, and the effect of blood transfusions. J. exp. Med. **90**, 97—111 (1947).

— The inheritance of vascular hemophilia: a new and interesting problem in human genetics. J. Med. Educ. **34**, 385—396 (1959).

GRETHER, W. F.: Color vision and color blindness in monkeys. Comp. Psychol. Monographs 15 (4), # 76 (1939).

GREWAL, M. S.: A sex-linked anemia in the mouse. Genet. Res. 3, 238—247 (1962).

GRÜNEBERG, H.: Genes and genotypes affecting the teeth of the mouse. J. Embryol. exp. Morphol. 14, 137—160 (1965).

HAMILTON, W. F., and T. B. COLEMAN: Trichromatic vision in the pigeon as illustrated by the spectral hue discrimination curve. J. comp. Psychol. 15, 183—191 (1933).

HSU, T. C., H. H. REARDEN, and G. F. LUQUETTE: Karyological studies of nine species of *Felidae*. Amer. Naturalist XCVII, 225—234 (1963).

HUTT, F. B., C. G. RICKARD, and R. A. FIELD: Sex-linked hemophilia in dogs. J. Hered. 39, 2—9 (1948).

JACOB, F., and J. MONOD: Genetic regulatory mechanism in the synthesis of proteins. J. molec. Biol. 3, 318—356 (1961).

KAZAZIAN, H. H. JR., W. J. YOUNG, and B. CHILDS: X-linked 6-phosphogluconate dehydrogenase in *Drosophila:* Subunit associations. Science 150, 1601—1602 (1965).

KIRKMAN, H. N., H. D. RILEY JR., and B. B. CROWELL: Different enzymic expressions of mutants of human glucose-6-phosphate dehydrogenase. Proc. Natl. Acad. Sci. U.S. 46, 938—944 (1960).

—, and E. M. HENDRICKSON: Glucose-6-phosphate dehydrogenase from human erythrocytes. II. Subactive states of the enzyme from normal persons. J. Biol. Chem. 237, 2371—2376 (1962).

— — Sex-linked electrophoretic difference in glucose-6-phosphate dehydrogenase. Amer. J. hum. Genet. 15, 241—258 (1963).

KLINGER, H. P.: The somatic chromosomes of some primates *(Tupaia glis, Nycticebus coucang, Tarsius bancanus, Cercocebus aterrimus, Symphalangus syndactylus)*. Cytogenetics 2, 140—151 (1963).

LADD-FRANKLIN, C.: Colour and colour theories. New York: Harcourt Brace & Co. 1929.

LEVAN, A., T. C. HSU, and H. F. STICH: The idiogram of the mouse. Hereditas 48, 676—687 (1962).

MANN, J. D., A. CAHAN, A. G. GELB, N. FISHER, J. HAMPER, P. TIPPETT, R. SANGER, and R. R. RACE: A sex-linked blood group. Lancet I, 8—10 (1962).

MARKS, P. A., and R. T. GROSS: Erythrocyte glucose-6-phosphate dehydrogenase deficiency; evidence of difference between Negroes and Caucasians with respect to this genetically determined trait. J. clin. Invest. 38, 2253—2262 (1959).

MATHAI, C. K., S. OHNO, and E. BEUTLER: Sex-linkage of the glucose-6-phosphate dehydrogenase gene in the family *Equidae*. Nature 209, 115—116 (1966).

MATTHEY, R.: La formule chromosomique et le problème de la détermination sexuelle chez *Ellobius lutescens* Thomas. *Rodentia-Muridae-Microtinae*. Arch. Klaus-Stift. Vererb.-L. 28, 65—73 (1953).

Matthey, R.: Un nouveau type de détermination chromosomique de sexe chez les mammifères *Ellobious lutescens* Th. et *Microtus (Chilotus) oregoni* Bachm. (Muridés-Microtinés). Experientia XIV/7, 240 (1958).

McKusick, V. A.: On the X-chromosome of man. Quart. Rev. Biol. **37**, 69—175 (1962).

Moore, K. L., and M. L. Barr: Morphology of the nerve cell nuclei in mammals, with special reference to the sex chromatin. J. comp. Neurol. **98**, 213—231 (1953).

Muhrer, M. E., A. G. Hogan, and R. Bogart: Defect in coagulation mechanisms of swine blood. Amer. J. Physiol. **136**, 355—359 (1942).

Mustard, J. F., H. C. Roswell, G. A. Robinson, T. D. Hoeksema, and H. G. Downie: Canine hemophilia B (Christmas Disease). Brit. J. Haemat. **6**, 259—266 (1960).

Nossel, H. L., R. K. Archer, and R. G. Macfarlane: Equine haemophilia: Report of a case and its response to multiple infusions of heterospecific AHG. Brit. J. Haematol. **8**, 335—342 (1962).

Ohno, S., W. D. Kaplan, and R. Kinosita: Formation of the sex chromatin by a single X-chromosome in liver cells of *Rattus norvegicus*. Exp. Cell Res. **18**, 415—418 (1959).

—, J. Trujillo, C. Stenius, L. C. Christian, and R. Teplitz: Possible germ cell chimeras among newborn dizygotic twin calves *(Bos taurus)*. Cytogenetics **1**, 258—265 (1962).

—, J. Jainchill, and C. Stenius: The creeping vole *(Microtus oregoni)* as a gonosomic mosaic. I. The OY/XY constitution of the male. Cytogenetics **2**, 232—239 (1963).

—, W. Beçak, and M. L. Beçak: X-autosome ratio and the behavior pattern of individual X-chromosomes in placental mammals. Chromosoma **15**, 14—30 (1964).

—, J. Poole, and I. Gustavsson: Sex-linkage of erythrocyte glucose-6-phosphate dehydrogenase in two species of wild hares. Science **150**, 1737—1738 (1965).

—, H. Payne, M. Morrison, and E. Beutler: Hexose-6-phosphate dehydrogenase found in human liver. Science **153**, 1015—1016 (1966).

Opitz, J. M.: Introduction to medical genetics. Part I. J. Iowa St. med. Soc. **51**, 393—409 (1961).

Parsons, J. H.: An introduction to the study of color vision. Cambridge (England): Univ. Press 1924.

Robinson, A. (ed.) (Denver Study group): A proposed standard system of nomenclature of human mitotic chromosomes. J. Amer. med. Ass. **174**, 159—162 (1960).

Sasaki, M. S., and S. Makino: Revised study of the chromosomes of domestic cattle and the horse. J. Hered. **53**, 157—162 (1962).

Shaw, C. R.: Electrophoretic variation in enzymes. Science **149**, 936—943 (1965).

—, and E. Barto: Autosomally determined polymorphism of glucose-6-phosphate dehydrogenase in *Peromyscus*. Science **148**, 1099—1100 (1965).

SHAW, C. R.: Glucose-6-phosphate dehydrogenase: Homologous molecules in deer mouse and man. Science 153, 1013—1015 (1966).

STEINBERG, A. G.: Progress in the study of genetically determined human gamma globulin types (the Gm and Inv groups). Progr. med. Genetics 2, 1—33 (1962).

TRUJILLO, J. M., C. STENIUS, L. C. CHRISTIAN, and S. OHNO: Chromosomes of the horse, the donkey and the mule. Chromosoma (Berl.) 13, 243—248 (1962).

—, B. WALDEN, P. O'NEIL, and H. B. ANSTALL: Sex-linkage of glucose-6-phosphate dehydrogenase in the horse and donkey. Science 148, 1603—1604 (1965).

WARNER, L. H.: The problem of color vision in fishes. Quart. Rev. Biol. 6, 329—348 (1931).

WOJTUSIAK, R. J.: Über den Farbensinn der Schildkröten. Z. vergl. Physiol. 18, 393—436 (1933).

YOUNG, W., J. E. PORTER, and B. CHILDS: Glucose-6-phosphate dehydrogenase in Drosophila: X-linked electrophoretic variants. Science 143, 140 (1964).

Chapter 5

Conservation of the Original Z-Chromosome by Diverse Avian Species and Homology of the Z-linked Genes

As stated in Chapter 3, the class *Aves* as a whole and the reptilian order *Squamata* appear to belong to the same genome lineage. Together they constitute one uniform group with the similar DNA value. The common characteristic shared by birds, snakes, and lizards is the possession of microchromosomes. Furthermore, it has been shown that the female heterogamety of the ZZ/ZW-type operates in snakes as well as in birds. Cytological evidence presented below suggests that the Z-chromosome of this genome lineage is very ancient in its origin. It appears that the same primitive Z has persisted in its entirety not only in diverse avian species, but also in diverse ophidian species of today.

a) Cytological Evidence of Conservation

Utilizing the same method which was applied for the measurement of the mammalian X, the absolute size of the Z-chromosome was estimated in seven avian species representing five different orders of diverse geographic origin (OHNO et al., 1964). Those studied were the canary *(Serinus canarius,* 2n = 80 ±) of the family *Fringillidae,* the order *Passeriformes;* the Australian parakeet *(Melopsittacus undulatus,* 2n = 58 ±) of *Psittacidae, Psittaciformes;* the pigeon *(Columba livia domestica,* 2n = 80 ±) of *Columbidae, Columbiformes;* and the duck *(Anas platyrhyncha domestica,* 2n = 80 ±) of the family *Anatidae,* the order *Anseriformes.* In addition, the order *Galliformes* was represented by three species belonging to two different families: the chicken *(Gallus gallus,* 2n = 78 ±) and the Japanese quail *(Coturnix coturnix japonica,* 2n = 78 ±) of the family *Phasianidae,* and the turkey *(Meleagris gallopavo,* 2n = 80 ±) of the family *Meleagridae.*

The karyotypes of the canary and the parakeet have already been shown in Fig. 7. The karyotypes of the other five species are shown in Fig. 15. With the exception of the parakeet, all of the birds in this series had a diploid number of about 80 with complements consisting of nine pairs of macrochromosomes, and about 30 pairs of microchromosomes which comprised 22 to 30% of the genome. The parakeet had a diploid number close to 60 (YAMASHINA, 1951; VAN BRINK, 1959) with 12 pairs of macrochromosomes and about 18 pairs of microchromosomes which comprised only 15% of the genome.

In all the species studied, the Z-chromosome could be singled out with absolute certainty. In the chicken, the Japanese quail, the turkey, the canary, and the pigeon, the Z-chromosome was a mediocentric element ranking either fourth or fifth largest in size. Autosomes of a similar size were either acrocentrics or subterminals, presenting little danger of confusing them with the Z. In the parakeet, the Z-chromosome was a submediocentric element fifth in the serial alignment, while the fourth pair was similar in size, but was composed of mediocentrics. In the duck, the Z was an acrocentric with acrocentrics on either side of it. Six acrocentrics roughly similar in size, comprising the fourth, fifth, and sixth pairs were found in the homogametic male, while the female had only five acrocentrics of this size range. In about half of the metaphase figures examined, the Z could be distinguished from the fifth and sixth pairs with certainty because of its prominent short arm and noticeably larger size (YAMASHINA, 1941).

Observation of Figs. 7 and 15 leaves the visual impression that the Z-chromosomes of seven avian species are nearly identical in size. The actual measurements confirmed the visual impression. The absolute size of the Z-chromosome was uniform, ranging only from 2.3 μ^2

in the pigeon and duck to 3.0 μ² in the canary. The Z comprised 7.20 to 9.32% of the genome. Thus, in the case of the class *Aves*, too, the cytological evidence supported the notion that the original Z-chromo-

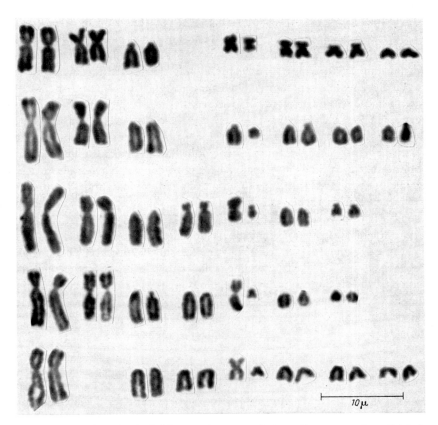

Fig. 15. The female diploid complements of five diverse species of birds demonstrating the extreme similarity in absolute size of the Z-chromosome. The Z is the fourth largest element in the pigeon, the duck and the turkey, while the Z is the fifth largest element in the chicken and the Japanese quail. Of these five species, the W can be singled out with absolute certainty only in the pigeon. Only the first seven pairs of macrochromosomes are aligned. Top row: The domestic pigeon (*Columba livia domestica*) representing the order *Columbiformes*. Second row: The domestic duck (*Anas platyrhyncha domestica*) of the order *Anseriformes*. Third row: The chicken (*Gallus gallus domesticus*) belonging to the order *Galliformes*. The remaining two also belong to the same order. Fourth row: The Japanese quail (*Coturnix coturnix japonica*). Fifth row: The turkey (*Meleagris gallopavo*)

some of a common ancestor has been preserved in its entirety by diverse avian species of today.

In various snake species constituting the reptilian suborder *Serpentes,* the uniformity in absolute size of the Z-chromosome was obvious. As stated in Chapter 2 and illustrated in Fig. 3, a great majority of snakes demonstrated very similar diploid complements made of eight pairs of macrochromosomes and ten pairs of microchromosomes. In every species, the Z was the fourth largest element having a median centromere. In every instance, the actual measurements have shown that the Z comprised little less than 10% of the genome (BEÇAK et al., 1964).

The most striking thing was the close similarity in absolute size of the avian Z-chromosome and the ophidian Z-chromosome. In view of the fact that birds and snakes belong to the same genome lineage, this close similarity appears to be more than coincidental. We prefer to interpret this fact as an indication that the selection of a specific homologous pair of autosomes as a future ZW-pair was already made in an ancient reptile ancestral to both snakes and birds which existed in the early part of the Mesozoic era (about 200 million years ago), and that the Z of this genome lineage has been preserved in its entirety to this date. Not only homology of the Z-linked genes among diverse avian species is expected, but it also appears probable that birds and snakes share the same kinds of Z-linked genes.

b) Homology of the Z-linked Genes among Diverse Species of Birds

As much as the Z-chromosome of birds comprises nearly 10% of the genome, the relative abundance of the known Z-linked genes is expected in the class *Aves.* Indeed, in such relatively well studied species as the chicken *(Gallus gallus domesticus)* (HUTT, 1949), the pigeon *(Columba livia domestica)* (LEVI, 1951), and the canary *(Serinus canarius)* (DUNCKER, 1928), a considerable number of Z-linked genes have been described. Unfortunately, all of the known Z-linked traits are morphological in nature, in most instances affecting the color pattern of the plumage. Since the direct product of a gene involved in a particular trait cannot be identified, possible homology between the Z-linked genes of different species has to be assessed on the basis of extreme similarity of the gene's phenotypic expression. Nevertheless, searching the literature for homologous Z-linked genes has been quite rewarding.

It appears that the Z-chromosome of diverse avian species carries a few gene loci which are concerned with the formation and distribution of melanin granules. A series of alleles which intensify and dilute in varying degrees the coloring of feathers has been found on each of these gene loci. At one extreme, a nearly amorphic allele gives an albino phenotype to the homozygous male as well as to the hemizygous female. The Z-linked albinism differs from the autosomal albinism in that the affected bird demonstrates residual pigmentation of the iris of the eye and exhibits light buff colored "ghost spotting" which imitates the normal pattern of colored feathers.

Apparently BATESON (1913) was already aware of the fact that whiteness of the peace dove *(Streptopelia risoria)* is due to the homozygous or hemizygous state of a Z-linked recessive gene. In a footnote on page 194 of his book *Mendel's Principle of Heredity,* he referred to the experiment carried out by WHITMAN in 1907. When the male *Streptopelia risoria* was crossed with the female ring dove *(Streptopelia humilis)* of dark brown color, whites in the F_1's were all females. The iris of the eye of *Streptopelia risoria* is pigmented, and a characteristic dark ring around the neck of *S. humilis* is seen as a ghost ring in *S. risoria.* Among other species belonging to the order *Columbiformes,* the Z-linked alleles *Almond* (B^{st}) and *Faded* (B^{of}) of the pigeon also give a practically white appearance with colored eyes to the homozygous males.

Sex-linked albinism has also been found in the Australian parakeet (STEINER, 1932) of the order *Psittaciformes.* Among those belonging to the order *Galliformes,* the sex-linked albinism has been found in the chicken, (HUTT, 1949) the Japanese quail (LAUBER, 1964) (Fig. 16), and the turkey (HUTT and MUELLER, 1942).

In the goose *(Anser domesticus)* belonging to the order *Anseriformes,* JEROME (1953) found that whiteness of the Emden breed is due to the combined action between the Z-linked dominant dilution gene *(Sd)* and an autosomal recessive spotting gene *(sp),* while whiteness of the White Chinese breed is due to the homozygous state of an autosomal recessive gene *(c).* The Z-linked dilution gene *(Sd)* of the chicken appears to be homologous with that found in the Emden breed of the goose. In the chicken, the *Sd* behaves like an allele of the Barred *(B)* which gives a characteristic black and white banding pattern to the feathers of the Barred Plymouth Rock breed. The males homozygous for *Sd* are white. The heterozygous male, having one Z-chromosome carrying *Sd* and the other Z carrying *B,* on the other hand, manifests dilution of black bands which are paled to indigo color.

Fig. 16. An appearance of a male Japanese quail homozygous for the sex-linked albino gene (right) is compared with that of a wild-type male (left). Light buff colored "ghost spotting" which imitates the wild-type plumage can be noted on an albino quail

In the canary of the order *Passeriformes,* the Z-linked gene *Isabell* inhibits the formation of eumelanin (black or brown), while not affecting the formation of phaeomelanin (yellow) granules. The result is the cinnamon colored bird with diluted pinkish eyes (DUNCKER, 1928). This is yet another evidence that the avian Z-chromosome carries the gene loci concerned with the formation and distribution of melanin granules.

From the above it may be concluded that the available genetic evidence, although indirect, tends to favor the notion that in the class *Aves,* too, the original Z-chromosome of a common ancestor has been conserved in its entirety by diverse species of modern birds.

c) Possible Homology between the Avian Z-linked Genes and the Ophidian Z-linked Genes

Because both birds and snakes demonstrate a nearly identical DNA value, and both the avian Z-chromosome and the ophidian Z comprise about 10% of the genome, it seems worthwhile to explore the possibility of homology between various Z-linked genes of birds and snakes. Unfortunately, we have not been able to find any report on the sex-linked hereditary trait of snakes. Although the extremely diluted phenotype which closely resembles the Z-linked albinism of birds has often been found in various ophidian species, to the best of my knowledge, no breeding test has been performed to determine the possible sex-linkage of this trait.

For the future, it would be particularly rewarding to find the Z-linked structural gene locus for an enzyme with multiple alleles in snakes and birds, each allele producing a distinct electrophoretic variant.

As stated in Chapter 2 and illustrated in Fig. 3, ophidian species belonging to the family *Boidae* are endowed with those sex chromosomes which are still in the primitive state of differentiation; the Z and the W are homomorphic. It is expected that, in these species, the W-chromosome should carry an allele of each Z-linked gene. If the W-linked allele produces a unique variant of an enzyme which is not produced by any of the alleles at the Z-linked locus, it would show that although the Z and the W are still largely homologous, the effective isolation between the two has been in operation for a considerable period of time. The study of the same enzyme on avian species and on snakes belonging to the family *Colubridae* and *Crotalidae* should enable us to determine the time when the W-linked locus allelic to the Z-linked gene disappeared in the differentiation process of the W as a result of genetic deterioration which occurred to the W.

d) Possible Homology between the Mammalian X-linked Genes and the Avian Z-linked Genes

Each diploid nucleus of placental mammals contains 7.0×10^{-9} mg DNA, the X-chromosome comprising about 5% of the genome. The DNA value for avian species, on the other hand, is 3.5×10^{-9} mg, the Z-chromosome comprising nearly 10% of the genome. The mammalian X and the avian Z are rather similar in absolute size. One may wonder if this similarity is a mere coincidence.

As stated earlier, the sex-linked albinism which is produced by a nearly amorphic allele at the particular Z-linked gene locus occurs to diverse species of birds. It is a curious fact that some of the known

X-linked mutations in the mouse *(Mus musculus)* can be regarded as homologous to the sex-linked albinism of the bird. In man, the sex-linked albinism with colored eyes has been reported.

In all the X-linked mutations belonging to the mottled series of the mouse (Table 1), only the heterozygous female survives; the affected hemizygous male invariably dies *in utero*. However, when the dead hemizygous male fetus is examined, it is found that the male is completely whitish, but has colored eyes (LYON, 1960). The variegated phenotype (white and light-colored spots on the background of wild-type) invariably manifested by the heterozygous female of the mottled series is the result of the peculiar dosage compensation mechanism which operates on the X-linked genes of mammals (see Part II). Excluding this peculiarity in the heterozygous state and the fact that the mutant gene is lethal in the hemizygous state, the mottled series of the mouse might be homologous with the sex-linked dilution series of the bird.

In man, the sex-linked albinism with slightly diluted eye colors has been found in 14 male descendants of one family in Alexandria, Egypt. The defect was apparently due to a recessive X-linked gene, as heterozygous females were apparently unaffected (ZIPRKOWSKI et al., 1962).

It is possible that both the mammalian X and the avian Z carry the same series of gene loci which are concerned with a particular phase of the melanin granule formation.

We became interested in seeking the possible Z-linkage of the structural gene for an enzyme, glucose-6-phosphate dehydrogenase in various avian species. Unfortunately, we were unable to find the intraspecific allelic polymorphism of G-6-PD in the chicken, Japanese quail, and pigeon. Interspecific hybrids such as the cross between the chicken and the Japanese quail were of no use as the chromosome study revealed that all viable hybrids were the genetic male of the ZZ-constitution.

In the future, if it is shown that the mammalian X and the avian as well as ophidian Z do share the same kinds of sex-linked genes, it would indicate that the decision in singling out a particular autosomal pair of a common ancestor as a future sex pair was not a random one. Placental mammals, snakes, and birds belong to two different genome lineages; thus, homology between the mammalian X-linked genes and the avian and ophidian Z-linked genes must indicate that the similar linkage group was independently preferred as the future sex chromosome.

References

BATESON, W.: Mendel's principles of heredity. Cambridge (England): University Press 1913.

BEÇAK, W., M. L. BEÇAK, H. R. S. NAZARETH, and S. OHNO: Close karyological kinship between the reptilian suborder *Serpentes* and the class *Aves*. Chromosoma (Berl.) **15**, 606—617 (1964).

DUNCKER, H.: Genetik der Kanarienvögel. Bibliographia genetica **9**, 37—140 (1928).

HUTT, F. B., and C. D. MUELLER: Sex-linked albinism in the turkey, *Meleagris gallopavo*. J. Hered. **33**, 69—77 (1942).

— Genetics of the fowl. New York: McGraw-Hill 1949.

JEROME, F. N.: Color inheritance in geese and its application to goose breeding. Poultry Sci. **32**, 159—165 (1953).

LAUBER, J. K.: Sex-linked albinism in the Japanese quail. Science **146**, 948—950 (1964).

LEVI, W. M.: The pigeon. Sex-linkage in pp. 243—249. Columbia, S. C.: The R. L. Bryan Comp. 1951.

LYON, M. F.: A further mutation of the mottled type in the house mouse. J. Hered. **51**, 116—121 (1960).

OHNO, S., C. STENIUS, L. C. CHRISTIAN, W. BEÇAK, and M. L. BEÇAK: Chromosomal uniformity in the avian subclass *Carinatae*. Chromosoma (Berl.) **15**, 280—288 (1964).

STEINER, H.: Vererbungsstudien am Wellensittich *Melopsittacus undulatus*. Arch. Klaus-Stift. Vererb.-L. **7**, 37—202 (1932).

VAN BRINK, J. M.: L'expression morphologique de la Digamétie chez les Sauropsidés et les Monotrèmes. Chromosoma **10**, 1—72 (1959).

YAMASHINA, Y.: Studies on sterility in hybrid birds. III. Cytological investigations of the intergeneric hybrid of the Muscovy duck *(Cairina moschata)* and the domestic duck *(Anas platyrhyncha* var. *domestica)*. Japan. J. Genetics **18**, 231—253 (1941).

— Studies on the chromosomes in twenty-five species of birds. Genetics **2**, 27—38 (1951).

ZIPRKOWSKI, L., A. KRAKOWSKI, A. ADAMS, H. COSTEFF, and J. SADE: Partial albinism and deaf-mutism due to a recessive sex-linked gene. Arch. Dermat. **86**, 530—539 (1962).

Part II

Evolution of Dosage Compensation Mechanism for Sex-linked Genes

Chapter 6

The Basic Difference in Constitution between the Mammalian X and the Drosophila X

During the course of evolution progressive genetic deterioration occurred to the Y-chromosome. As a result, the Y-chromosome of placental mammals became a highly specialized male-determiner, having but one function namely to induce the indifferent embryonic gonad to develop as testis. The consequence of this specialization by the Y was that almost all of the original Mendelian genes which had been maintained by the X-chromosome, had to accommodate themselves to a hemizygous existence in the heterogametic male sex. The hemizygous existence for all the genes on the X should be very perilous since monosomy for even the smallest autosome (only one-fourth the size of the X) is apparently lethal in man.

The fact that the male thrives almost as well as the female suggests that a buffer mechanism evolved, offsetting the genetic disparity between the male with only one X and the female with two X's. Based on the findings made on X-linked genes of *Drosophila melanogaster,* MULLER (1947) defined this buffer mechanism as the dosage compensation mechanism for individual X-linked genes.

As will be shown in the next chapter, the dosage compensation mechanism as applied to the X-linked genes of *Drosophila* is fundamentally different from that which operates on X-linked genes of placental mammals. The difference is clearly the reflection of the constitutional dissimilarity that exists between the *Drosophila* X and the mammalian X. Cytological observations which led to the discovery of this constitutional dissimilarity shall be elaborated in this chapter.

a) The Historical Contribution Made by the X-chromosome in the Formulation of the Heterochromatin Concept

Ten years before the discovery of the chromosomal sex-determining mechanism by McClung (1902), the X-chromosome as such had been recognized under the microscope, and its behavior in male germ cells described with exquisite clarity by Henking in 1891. In the testis of the heteropteran insect, *Pyrrhocoris apterus* (2n = 23/24, XO/XX), Henking observed that at first meiotic prophase in each nucleus of primary spermatocytes one densely stained round body stood out in sharp contrast to the fine, thread-like chromosomes. He regarded this body as a nucleolus. However, when he examined a pair of secondary spermatocytes, he noted that this densely stained body was always incorporated into one of the two daughter spermatocytes. Two kinds of secondary spermatocytes were produced in equal number: one with 11 chromosomes only, and the other with this particular body in addition to the 11 chromosomes. He was no longer certain that this body was a mere nucleolus, and since he arrived at no clear conclusion, he labeled it "X" for unknown. It is in Henking's honor that today the term X-chromosome denotes the sex chromosome of the male heterogamety which is associated with the development of the female sex. This historical episode is used to show that the well differentiated X-chromosomes seen in insects among invertebrates, and mammals among vertebrates, have an inherent tendency to remain condensed during interphase and prophase while other chromosomes elongate.

As chromosome cytology became more sophisticated, Heitz (1933) defined those chromosomal regions which stain darkly and remain condensed to form chromocenters in the interphase nucleus as heterochromatic regions. In contrast, the remaining chromosomal regions which comprise the fine network of chromatin in the interphase nucleus are said to be made of euchromatin.

b) The Euchromatic and Heterochromatic Regions of the Drosophila X

Heitz (1933) and Kaufmann (1934) found the X-chromosome of *Drosophila melanogaster* to be structurally divided. The proximal part, including the centromere and a nucleolus-organizer, is made of heterochromatin, while the distal part is made of euchromatin. As somatic pairing of homologous chromosomes occurs in dipteran insects, the heterochromatic regions of two X's of the female *Drosophila* are recognized in interphase nuclei of larval ganglion cells as closely paired chromocenters attached to the nucleolus (Fig. 17). Extensive parallel genetic analysis carried out on this species by MORGAN,

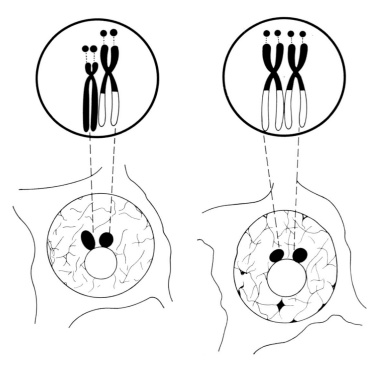

Fig. 17. Schematic representation of the X and the Y of the male and the two X's of the female of *Drosophila melanogaster* modeled after HEITZ (1933). The male is at the left and the female at the right; metaphase chromosomes at the top, and interphase nuclei of larval ganglion cells at the bottom. Heterochromatic regions of the sex chromosomes are painted solid black. Both the submediocentric X and the subterminal Y carry a nucleolus organizer. The Y is entirely heterochromatic, while the X is partially so. Somatic interphase nuclei of both sexes exhibit the two chromocenters derived from the sex chromosomes adjacent to a nucleolus. Because of somatic pairing, the two chromocenters are found in very close proximity of each other

STURTEVANT, MULLER, and STERN, among others, revealed that the heterochromatic region of the X is a genetically empty dummy for X-linked genes, and female-determining factors are found almost exclusively on the euchromatic region. The Y-chromosome of this species is entirely heterochromatic; neither the male-determining factors nor the Mendelian genes are found on the Y (the gene *bobbed*, which shall be mentioned in the next chapter, is an exception). It then seemed that the heterochromatin was inherently different from the euchromatin; hence, the concept "once heterochromatin, always heterochromatin".

c) Heterochromatinization not as the Consequence of Genetic Deterioration, but as the Functional State Assumed by Individual X-chromosomes

Little was it realized by *Drosophila* geneticists that even among insects the X-chromosome of *Drosophila melanogaster* is exceptional. HENKING's original observation (1891) had revealed that in male germ cells of *Pyrrhocoris apterus*, the X, condensed along its entire length, was not segmented into heterochromatic and euchromatic regions. Furthermore, the concept of "once heterochromatin, always heterochromatin" does not apply to the X of most insects, a point vividly illustrated in 1923 by JUNKER's observation on X_1X_2O-hermaphrodites of the pleicopteran insect *Perla marginata* (2n = 22) (JUNKER, 1923). In the testicular part of the ovotestis, both the X_1 and the X_2 in primary spermatocytes appeared heterochromatic along their entire lengths throughout first meiotic prophase, while the same X_1 and X_2-chromosomes in the ovarian part were behaving as euchromatin throughout all stages of first meiotic prophase of oöcytes. Thus, under certain circumstances, an individual X-chromosome appears to be entirely heterochromatic, but this is the functional state temporarily assumed by the X and does not reflect genetic emptiness.

During meiosis of the heterogametic male sex, the X of most mammals, including man, behaves in exactly the same manner as the X of the heteropteran insect. At pachytene, the sex vesicle (a densely stained round body formed by the X and the Y) is easily distinguished from the thread-like autosomal bivalents with their intricate chromomeric patterns. Throughout interkinesis and second prophase of secondary spermatocytes, the X or the Y stands out from the haploid set of autosomes in each nucleus by virtue of their positive hetero-pyknosis (TJIO and LEVAN, 1956; OHNO et al., 1959 a).

The XX-bivalent of female mammals, on the contrary, demon-strates as fine a chromomeric pattern as any of the autosomal bi-valents at pachytene (OHNO et al., 1961). Even the X-univalent in oöcytes of the XO-mouse does not manifest the heterochromatic con-dition, but instead behaves in an euchromatic manner (JAGIELLO and OHNO, 1966).

The mammalian X is entirely heterochromatic during male meiosis and entirely euchromatic during female meiosis. The clue to the unique behavior pattern of individual X-chromosomes in somatic cells of placental mammals lies in this ambivalent nature of the mammalian X.

d) The Single X-derivation of the Barr Sex Chromatin Body of Mammalian Female Somatic Cells

A distinct sexual dimorphism of mammalian somatic interphase nuclei was first noted in the cat by BARR and BERTRAM in 1949. While observing nuclei of nerve neurons in histological sections, they noted that individual nuclei of the female almost invariably contained a distinct chromocenter (a body of heterochromatin) of a constant size adjacent either to a nucleolus or to the nuclear membrane. The comparable chromocenter was never found in nerve neurons of the male cat (BARR and BERTRAM, 1949). Subsequently, BARR and his colleagues established that this sexual dimorphism occurs to most somatic cell types of a great majority of mammals, including man. Thus, this female-specific chromocenter became known as the "Barr sex chromatin body".

Because mammalian cytology and genetics were still in their infancy in 1949, the nature of the Barr sex chromatin body was not immediately understood. Accordingly, workers naturally sought a model in *Drosophila melanogaster*. The concept of "once hetero-chromatin, always heterochromatin" appeared unchallengeable. It was assumed that, in mammals too, the X must be made of one part heterochromatin and one part euchromatin. A bipartite structure occasionally revealed by the sex chromatin body (KLINGER, 1958) appeared to substantiate the interpretation that the female-specific chromocenter of mammals represented fused heterochromatic regions of two X's. It was indeed comforting to know that this view was in complete accord with what was known in *Drosophila*. The observation that the male somatic nucleus lacked the half-sized sex chromatin

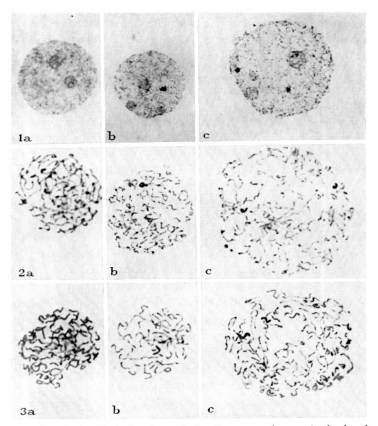

Fig. 18. The single X derivation of the Barr sex chromatin body shown by photomicrographs of regenerating liver cells of the rat *(Rattus norvegicus)*, from our original paper of 1959 (OHNO et al., 1959 b). Row (1) demonstrates interphase nuclei; row (2), early prophase and row (3), middle prophase figures. Column (a) shows male diploid; column (b), female diploid and column (c), female tetraploid nuclei. The male interphase nucleus is sex-chromatin-negative (1 a). At prophase, all the chromosomes appear in a fine thread-like state (2 a, 3 a). The female diploid nucleus has a single sex-chromatin body (1 b) which is visualized as a single chromosome condensed along its entire length in prophase nuclei (2 b, 3 b). The female tetraploid has two sex chromatin bodies (1 c). Accordingly, there are two entirely condensed chromosomes in each tetraploid prophase (2 c, 3 c)

body representing a heterochromatic region of a single X was dismissed as a mere technical difficulty.

As it turned out, the sex chromatin body of female somatic interphase nuclei was found to represent a single X-chromosome in an entirely heterochromatic state. In 1959, we examined squash preparations of regenerating liver of the rat *(Rattus norvegicus,* 2n = 42). While observing diploid prophase figures of the female, it was noted that the sex chromatin body of the preceding interphase was not resolved as heterochromatic regions of two chromosomes in somatic pairing. Instead, the presence of a rather large chromosome heavily condensed along its entire length was conspicuous in each diploid prophase. Nearly 20% of the interphase nuclei contained in the squash preparations from the female were apparently tetraploid, demonstrating two sex chromatin bodies in each nucleus. Accordingly, each female tetraploid prophase figure contained two such heavily condensed chromosomes. In sharp contrast, neither diploid nor tetraploid prophase nuclei of the male included a condensed chromosome of comparable size. All the chromosomes of the male appeared in a thread-like, extended state at prophase (Fig. 18). On the basis of the above observation, we have concluded that each sex chromatin body actually represents a single X-chromosome which manifests the positively heteropyknotic condition along its entire length. Despite the single X-derivation of the sex chromatin body, a sexual dimorphism occurs because the X-chromosome of the male, as well as one of the two X's of the female, invariably behaves in an euchromatic or isopyknotic manner, assuming the finely dispersed state during interphase as do the autosomes (OHNO et al., 1959 b).

In the mouse *(Mus musculus,* 2n = 40), a sexual dimorphism is not at all evident in the interphase nuclei. The presence of several chromocenters (each probably representing a fused mass of several autosomal centromeric heterochromatin) in both sexes obscures the exclusive occurrence in the female of the sex chromatin body. Yet, at prophase, one of the two X-chromosomes in a somatic cell of the female mouse was as clearly heterochromatic as that found in the female rat (OHNO and HAUSCHKA, 1960). The single X-derivation of the sex chromatin body of man was also confirmed at prophase (OHNO and MAKINO, 1961).

Using tritiated thymidine as a label, TAYLOR (1960) studied the DNA replication pattern of individual chromosomes in somatic cells of the Chinese hamster *(Cricetulus griseus,* 2n = 22). He found that the heterochromatic X of the female replicates its DNA late in the synthetic phase of interphase, after the autosomes and the euchromatic X have nearly completed their replication. Subsequently, it was confirmed in man (Grumbach and Morishima, 1962; German, 1962; Gilbert et al., 1962), the mouse (Galton and Holt, 1965), and many other placental mammals that the late replication of DNA is the characteristic of an individual X-chromosome which behaves in the heterochromatic manner.

e) Brief Summary of the Constitutional Difference between the Drosophila X and the Mammalian X

In the case of the X-chromosome of the fruit fly *(Drosophila melanogaster)*, the heterochromatic and euchromatic regions of the X represent two inherently different segments of the X. Extensive genetic deterioration occurred to the heterochromatic region. As a result, it is now genetically empty, and it invariably forms an interphase chromocenter. All the X-linked genes and female-determining factors are located on the euchromatic region. This euchromatic region never assumes the heterochromatic state.

The original-type X of placental mammals which comprises 5% of the genome, on the other hand, is not segmented into two regions. The indications are that the X-linked genes and female-determining factors are distributed along the entire length of the X. The heterochromatic and euchromatic states are the two functional states which can be assumed by individual X-chromosomes under different circumstances. In somatic cells of the male a single X always behaves in an euchromatic manner. In somatic cells of the female, one X-chromosome still behaves in an euchromatic manner while the other X invariably becomes entirely heterochromatic and forms the Barr sex chromatin body. So far as somatic cells are concerned, both the XY-male and the XX-female have only one euchromatic X in each nucleus. If the euchromatic state can be correlated with the active state, and the heterochromatic state with the inert state, both sexes would have but one functioning X in each somatic nucleus. It is easy to see that in mammals this might be the means to achieve the dosage compensation of X-linked genes.

References

BARR, M. L., and L. F. BERTRAM: A morphological distinction between neurones of the male and female and the behavior of the nucleolar satellite during accelerated nucleoprotein synthesis. Nature **163**, 676—677 (1949).

GALTON, M., and S. F. HOLT: Asynchronous replication of the mouse sex chromosomes. Exp. Cell Res. **37**, 111—116 (1965).

GERMAN, J. L.: DNA synthesis in human chromosomes. Trans. N. Y. Acad. Sci. **24**, 395—407 (1962).

GILBERT, C. W., S. MULDAL, L. G. LAJTHA, and J. ROWLEY: Time-sequence of human chromosome duplication. Nature **195**, 869—873 (1962).

GRUMBACH, M. M., and A. MORISHIMA: Sex chromatin and the sex chromosomes: on the origin of sex chromatin from a single X-chromosome. Acta Cytol. **6**, 46—60 (1962).

HEITZ, E.: Die somatische Heteropyknose bei *Drosophila melanogaster* und ihre genetische Bedeutung. Z. Zellforsch. Abt. Histochem. **20**, 237—287 (1933).

HENKING, H.: Untersuchungen über die ersten Entwicklungsvorgänge in den Eiern der Insekten. II. Über Spermatogenese und deren Beziehung zur Eientwicklung bei *Pyrrhocoris apterus* L. Z. wiss. Zool. **51**, 685—736 (1891).

JAGIELLO, G., and S. OHNO: Isopycnotic behavior of the X-univalent in the XO mouse ovum. Exp. Cell Res. **41**, 459—462 (1966).

JUNKER, H.: Cytologische Untersuchungen an den Geschlechtsorganen der halbzwitterigen Steinfliege *Perla marginata* (Panzer). Arch. Zellforsch. **17**, 185—259 (1923).

KAUFMANN, B. P.: Somatic mitoses of *Drosophila melanogaster*. J. Morphol. **56**, 125—156 (1934).

KLINGER, H. P.: The fine structure of the sex chromatin body. Exp. Cell Res. **14**, 207—211 (1958).

McCLUNG, C. E.: The spermatocyte divisions of the *Locustidae*. Kansas Univ. Sci. Bull. **1**, 185—238 (1902).

MULLER, H. J.: Evidence of the precision of genetic adaptation. Harvey Lectures Ser. **43**, 165—229 (1947—1948).

OHNO, S., W. D. KAPLAN, and R. KINOSITA: On the end-to-end association of the X and Y-chromosomes of *Mus musculus*. Exp. Cell Res. **18**, 282—290 (1959 a).

— — — Formation of the sex chromatin by a single X-chromosome in liver cells of *Rattus norvegicus*. Exp. Cell Res. **18**, 415—418 (1959 b).

— — — X-chromosome behavior in germ and somatic cells of *Rattus norvegicus*. Exp. Cell Res. **22**, 535—544 (1961).

—, and T. S. HAUSCHKA: Allocycly of the X-chromosome in tumors and normal tissues. Cancer Res. **20**, 541—545 (1960).

—, and S. MAKINO: The single X nature of the sex chromatin in man. Lancet I, 78—79 (1961).

TAYLOR, J. H.: Asynchronous duplication of chromosomes in cultured cells of Chinese hamster. J. Biophys. Biochem. Cytol. 7, 455—464 (1960).

TJIO, J. H., and A. LEVAN: Notes on the sex chromosomes of the rat during male meiosis. Anales Estac. Exp. Aula Dei 4, 173—184 (1956).

Chapter 7

The Two Different Means of Achieving Dosage Compensation for X-linked Genes Employed by Drosophila and Mammals

As shown in the previous chapter, the X-linked genes of *Drosophila* are included in the euchromatic region which never becomes heterochromatic. The male is endowed with one dose, while the female receives two doses of the euchromatic region of the X. During the course of evolution of this insect, the dosage compensation mechanism must have developed one by one for each individual X-linked gene.

The X-linked genes of placental mammals are distributed along the entire length of the original X. As a result of the sex chromatin body formation by one of the two X's of the female, both the male and the female are endowed with one euchromatic X in each somatic nucleus. It is likely that dosage compensation was accomplished for all the X-linked genes in one sweep by heterochromatinization of one of the two X's of the female.

In this chapter the dosage compensation mechanism which operates for the X-linked genes of *Drosophila* shall be compared with that which operates for the X-linked genes of placental mammals. When we speak of dosage compensation, it should be pointed out that a need for dosage compensation exists only for the Mendelian genes on the X (so-called X-linked genes). Such a need does not exist for the female-determining factors on the X. In fact, in the case of *Drosophila*, the sex-determining mechanism is dependent upon the dosage effect of the female-determining factors. The feminizing influence emanating from the single X is not sufficient to overpower the diploid set (2 A) of autosomes carrying a fixed dose of the male-determining factors. It requires the two X's to overcome the male-determining influence from the diploid set of autosomes (see Chapter XIII).

The Y-chromosome of placentals appears to be made entirely of heterochromatin as it invariably manifests the positively heteropyknotic condition and demonstrates late DNA replication. In part, this is a reflection of extensive genetic deterioration that occurred to the Y. All the Y-linked genes which were formerly allelic to the X-linked genes have long disappeared, yet the Y exerts the very strong male-determining influence. Obviously, the sex-determining factors differ from the ordinary Mendelian genes in that they are not inactivated by heterochromatinization. The same should hold true for the female-determining factors on the mammalian X. This chapter concerns only the dosage compensation for X-linked genes, and not for sex-determining factors.

a) The Gene Dosage Effect as the Basis for a Need to Compensate for a Single Dose in the Male of Each X-linked Gene

The X-chromosome exists alone in the male; consequently, the male is equipped with only a single dose of each X-linked gene. This does not seem too much of a peril until it is realized that the rate of product output by the structural gene is rigidly fixed either by the structural gene itself or by the operator gene adjacent to it. Thus, the amount of an enzyme or a non-enzymatic protein produced by a single dose of the structural gene is exactly half of that produced by double doses of the same structural gene.

This point shall be illustrated in a few well proven examples. Galactosemia of man is an autosomally inherited recessive trait. The cause is a defective mutation at the autosomal gene locus for an enzyme, galactose-1-phosphate uridyl transferase (ISSELBACHER et al., 1956). The enzyme level in erythrocytes of an affected individual homozygous for this defective mutant allele is zero. The enzyme level in erythrocytes of heterozygotes (both parents of an affected) is found to be 50% of the level found in normals (ROBINSON, 1963). The 50% reduction of the normal enzyme level also occurs in human individuals heterozygous for other autosomally inherited defective traits. The 50% enzyme level has been observed in the heterozygotes for acatalasemia (TAKAHARA et al., 1960; NISHIMURA et al., 1961) as well as for non-spherocytic congenital hemolytic anemia due to a deficiency of pyruvate kinase (TANAKA et al., 1962).

The above examples on heterozygotes having one functional wild-type allele and one defective amorphic allele at a structural gene locus clearly reveal the fact that the rate of synthesis of the product by individual diploid cells is dependent upon the number of functional genes contained in each nucleus. If there is a relatively greater contribution of the normal allele in heterozygotes as a means of compensation for the presence of a defective allele, enzyme levels higher than 50% are to be expected.

The fact that the rate of product output by each functional allele at the structural gene locus is rigidly fixed beyond the influence of changes in physiological environments, is more clearly illustrated by our recent study on the enzyme, 6-phosphogluconate dehydrogenase of Japanese quail *(Coturnix coturnix japonica)* and the chicken *(Gallus gallus domesticus)*. As stated in Chapter 4, 6-phosphogluconate dehydrogenase (6-PGD) catalyzes the second step of pentose phosphate shunt of carbohydrate metabolism. It converts 6-phosphogluconate to ribulose-5-phosphate. In placental mammals, glucose-6-phosphate dehydrogenase which catalyzes the first step of this shunt is produced by the X-linked gene, while 6-PGD is autosomally inherited (see Chapter 4).

In a stock of Japanese quails maintained at the Department of Poultry Husbandry, University of California at Davis by HANS ABPLANALP and LESLIE E. HALEY, three electrophoretic variants of 6-PGD were found, enabling us to elucidate the presence of a single, autosomally inherited gene locus for this enzyme in this species. While three homozygous types demonstrated a single sharp 6-PGD band of a characteristic mobility, each of the three heterozygous types revealed the coexistence of three forms of the enzyme. In addition to the two electrophoretic variants inherited from the parents, a new band of an intermediate mobility was seen in the middle. The 6-PGD molecule is a dimer having the molecular weight of about 117,000. When two different alleles are brought together in a heterozygote, a hybrid dimer molecule is produced in addition to two types of autodimers. This accounts for the three band pattern in heterozygotes. The heterozygous types occur with equal frequency in both sexes, and therefore the autosomal inheritance of this enzyme is well established (Fig. 19).

When the absolute amount of 6-PGD contained in individual erythrocytes is compared with the three homozygous types and the three heterozygous types by measuring the rate of reduction of TPN to TPNH, it is found that the six possible combinations of three different alleles produce exactly the same amount of the enzyme per unit of hemoglobin. The mean value of 2.8 is invariably obtained.

So far as we can determine, chickens of different breeds demonstrate no allelic polymorphism at the 6-PGD locus. The wild-type
gene of the chicken, however, is approximately 3.5 times more pro-
ductive than the three alleles of the quail. The 6-PGD activity per
unit of hemoglobin in the chicken is 9.2.

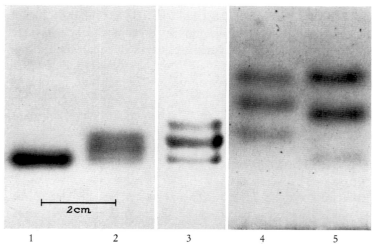

Fig. 19. A photograph of a starch gel plate stained for 6-phosphogluconate
dehydrogenase. Erythrocyte extracts of Japanese quails (Slots 1, 2 and 3)
and chicken-quail hybrids (Slots 4 and 5). Vertical starch gel electrophoresis
was carried out for 14 hours at pH 8.6 in borate buffer with a gradient of
4 volt/cm. 1) The male quail homozygous for the slowest moving C-allele.
2) The female quail heterozygous (B/C). A hybrid dimer band in the middle
is most intensely stained. This is also true of the (A/C) heterozygote. 3) The
female quail heterozygous (A/C). 4) The male hybrid which received the
B-allele from his quail mother. The chicken wild-type allele at the top is
most intensely stained. 5) The male hybrid which received the C-allele from
his quail mother

If each allele at the structural gene locus has the rigidly fixed rate
of the self-imposed product output, the interspecific hybrid between
the Japanese quail and the chicken should show the 6-PGD activity
value of 6.0, which is the sum of the productivity of one quail allele
and one chicken allele $\frac{2.8+9.2}{2}$. The actual value of 5.5 obtained on
these interspecific hybrids is indeed very close to the expected value.

This observation is meaningful only if each quail 6-PGD polypeptide is as enzymatically active as each chicken polypeptide, the

difference being in the rate of synthesis of polypeptides. If the above is the case, the ratio of three 6-PGD isozymes in the hybrid should be as follows: chicken dimer : hybrid dimer : quail dimer = 11 : 7 : 1 (Fig. 19). Indeed, our dilution experiment substantiated this expectation. If the difference is due to lesser enzymatic activity of each quail polypeptide, the expected ratio changes to 3:4:1.

Because the rate of the product output by each structural allele is rigidly fixed, one would expect that unless something has been done to remedy the situation, the G-6-PD level of the male mammals should be 50% that of the female, for G-6-PD of man and other placental mammals is produced by the X-linked structural gene. In fact, something has indeed been done. As mentioned in Chapter 4, an amorphic mutant allele at the G-6-PD gene locus occurs with a relatively high frequency among Mediterranean peoples. The erythrocyte enzyme level among the affected of both hemizygous males and homozygous females is almost zero. The heterozygous female with one dose of the normal functional allele demonstrates the expected 50% reduction of the enzyme level (ADINOLFI et al., 1960). Yet, the normal male with the same single dose of the normal allele maintains the 100% enzyme level as does the normal female with two doses of the same allele. Something has apparently been done to the X-linked gene loci for antihemophilic globulin (AHG) and plasma thromboplastic component (PTC) of man as well. There is no appreciable difference between the normal male and the normal female with regard to their blood AHG and PTC levels. Yet, the 50% reduction of the AHG level is seen in the female who is a heterozygous carrier of hemophilia A (RAPAPORT et al., 1960). The same occurs to the PTC level of a heterozygous female carrier of hemophilia B (BARROW et al., 1960).

b) Dosage Compensation in Drosophila through a Series of Modifier Genes for Individual X-linked Genes

In the early 1930's, great *Drosophila* geneticists such as H. J. MULLER and C. STERN realized that nature has done something to adjust the dosage effect of X-linked genes in this fly, long before it became known that each gene is a DNA molecule which specifies the amino acid sequence of a polypeptide it produces. MULLER (1947) defined this something as the dosage compensation mechanism for individual X-linked genes. The concept of dosage compensation as applied to *Drosophila* involves the existence of a modifier gene or genes within the X-chromosome which cancels the effect of different doses of a given structural gene. Thus, the male with a single dose of the mutant gene *apricot* demonstrates as much *apricot* eye color as the female with two doses of *apricot*. The degree of deepening of the eye color attainable in the female by having four doses of *apricot* can be attained in the male by only two doses of *apricot*.

The actual evidence in support of the postulate that one X-linked gene can act as a modifier gene of the other X-linked gene was supplied later by GANS (1953). Normally, the male fly with one dose of the X-linked mutant gene *zeste* (z) has normal red eyes while females with two doses of z have lemon colored *zeste* eyes. The dosage is not compensated for this X-linked gene. When an extra dose of the normal allele of *white* (w^+) is added to an X by a partial duplication, however, the male with still one dose of z now becomes *zeste* eyed. Conversely, when w^+ is deleted from one of the two X-chromosomes of the female, females with still two doses of z, but only one dose of w^+ are now equipped with normal red eyes. This shows that the X-linked gene w^+ can act as a modifier gene of the other X-linked gene z.

The *bobbed* is an exceptional X-linked gene of *Drosophila* in that the Y-chromosome also carries this gene. There is absolutely no need for dosage compensation for this gene because the XY-male and the XX-female both carry the two doses. In fact, STERN (1929) has shown that there is a definite dosage effect of *bobbed*. XO-males with their single dose of this gene have much shorter bristles than XY-males and XX-females with their two doses.

There are other X-linked genes of *Drosophila* which show either no sign or an insufficient sign of dosage compensation. For instance, the male hemizygous for *eosine* demonstrates lighter eye color than the female homozygous for *eosine*.

It appears that because the X-linked genes of *Drosophila* are located on the euchromatic region which always remains functional, the dosage compensation had to be accomplished in piecemeal fashion by developing a set of modifier genes for each structural gene. Thus, the evolutionally new X-linked genes which were recently transposed to the X from autosomes, may not have had sufficient time to develop their own modifier genes which would account for the lack of dosage compensation for certain X-linked genes of this species.

In essence, the dosage compensation mechanism of *Drosophila* appears to operate by forcing a given X-linked gene in the XY-male to work harder, while restraining the activity of the same X-linked gene on each of the two X's of the female. The recent study by KAZAZIAN and his colleagues (1965) on the enzyme 6-phosphogluconate dehydrogenase gave the direct confirmation to the dosage compensation mechanism postulated above. Unlike in mammals and birds, an X-linked structural gene locus controls the production of 6-PGD in *Drosophila*. The two alleles which produce electrophoretic variants A and B were found at this gene locus. When starch gel electrophoresis was performed on extracts from heterozygous (A/B) female flies, the coexistence of the three distinct enzyme bands was noted. The middle band representing a hybrid dimer enzyme was more intensely stained than the two autodimer bands, A_2 and B_2. The formation of the hybrid dimer molecules by heterozygotes requires that in each individual somatic cell A-subunit and B-subunit polypeptides are produced at the same rate. The two alleles at the X-linked 6-PGD gene locus of the female are functioning in synchronous harmony. Although the male fly with a single dose of the 6-PGD gene maintains the same enzymatic level as the female with the two doses, the dosage compensation is not accomplished by inactivation of one or the other alleles in the female. This is the fundamental difference that exists between the dosage compensation mechanism of *Drosophila* and that of placental mammals.

c) Dosage Compensation by X-inactivation in Placental Mammals

When the single X-derivation of the sex chromatin body came to light, it occurred to many of us that heterochromatinization of one of the two X's of the female might be the means for achieving the dosage compensation for all the X-linked genes at once. The heterochromatic state might be the reflection of functional inactivation. If this were the case, then both the male with one X and the female with two X's should actually have one functioning X in each somatic cell. The genetic disparity between the sexes would therefore be equalized.

The elucidation of the dosage compensation mechanism of mammals, however, we owe to MARY F. LYON (1961). Ingenuity of LYON's hypothesis lies in her assumption that random inactivation of one or the other X occurs early in the embryonic development of the female. When inactivation of one X-chromosome occurs early in the female's embryonic life, it is a random matter as to which X-chromosome in any single cell is affected. But once an X-chromosome has been inactivated, it remains inactivated in all descendants of that cell. Thus, the mammalian female is a natural mosaic with two distinct populations of somatic cells: one with a functioning paternally-derived X, and the other with a functioning maternally-derived X.

It comes as no surprise that the clearest support to this hypothesis has been furnished by the study of the X-borne enzyme, glucose-6-phosphate dehydrogenase (G-6-PD). Among numerous X-linked hereditary traits of man, the mouse, and other placental mammals which we know today, the structural gene for G-6-PD is probably the only X-linked gene which gives us a direct product that may be measured and analyzed.

Shortly after Lyon advanced her hypothesis, Beutler and his colleagues (1962) showed that women heterozygous for a defective mutant G-6-PD gene have two populations of erythrocytes: one produced by erythroblasts having the inactivated X carrying the defective mutant gene and showing the normal 100% enzyme level, the other produced by those having the inactivated normal X and showing a nearly zero enzyme level.

The dosage compensation by random inactivation of one or the other X of the female should result in the invariably hemizygous expression of the X-linked gene in each individual somatic cell of the mammalian female. If the female carries two different alleles at the G-6-PD gene locus, only one or the other allele should be expressing itself in each somatic cell. The above-mentioned study by Beutler and his colleagues on G-6-PD levels of erythrocytes appeared to confirm the hemizygous expression of X-linked genes in female somatic cells, but the most direct evidence was furnished by the cloning experiment carried out by Davidson and his colleagues (1963). As stated in Chapter 4 and illustrated in Figure 13 a, the wild-type, or species-specific G-6-PD of man is known as the electrophoretic variant B. In addition, the more negatively charged and therefore faster moving variant occurs with a high frequency among Americans of African descent. Some women are heterozygous, having one allele for the B-variant and the other allele for the A-variant. When an extract from several hundred million erythrocytes, leukocytes, or fibroblasts is subjected to starch gel electrophoresis, the coexistence of A- and B-variants can be visualized as two distinct bands with different electrophoretic mobilities.

DAVIDSON and his colleagues (1963) obtained skin biopsy material from six heterozygous (AB) women. Two were heterozygous for the functional A⁺-variant, while four were heterozygous for the hypomorphic A⁻-variant. From the fibroblast culture of each woman, several clones were developed by single cell platings. When starch gel electrophoresis was performed on an extract of each clone derived from a single somatic cell, the solitary existence of either the B-variant band or the A-variant band was noted. None of the 54 clones derived from the six heterozygous women demonstrated the coexistence of the A- and B-variants. Furthermore, the clone which demonstrated the A⁻-variant revealed the very low enzyme level comparable to that exhibited by the cultured fibroblasts from the hemizygous A⁻-male.

When this result obtained on G-6-PD of man by DAVIDSON and his colleagues is compared with the result obtained on 6-PGD of *Drosophila* by KAZAZIAN and his colleagues (1965), the fundamental difference that exists between the dosage compensation mechanism employed by mammals and that employed by *Drosophila* is clearly realized.

d) Evolutional Steps Followed by Ancestors of Placental Mammals which Led to the Development of the Dosage Mechanism by X-inactivation

During the course of evolution, an ancestor to placental mammals must have escaped a peril resulting from the hemizygous existence of all the X-linked genes in the male by doubling the rate of product output of each X-linked gene. Once this step was accomplished, the female no longer needed two X's in her somatic cells. Hence, the dosage compensation mechanism by random inactivation of one or the other X evolved.

In the case of *Drosophila*, on the other hand, it appears that a needed increase of the rate of product output by individual X-linked genes did not take place in their evolutional past. Thus, two alleles at each X-linked gene locus are still needed by the female. The presence of modifier genes is required primarily to raise the efficiency of individual X-linked genes in the hemizygous state as a means of minimizing a peril encountered by the male.

References

Adinolfi, M., I. Bernini, V. Carcassi, B. Latte, A. G. Motulski e M. Siniscalco: Indagine genetiche sulla predisposizione al favismo. Acad. nat. Lincei, R. C., **28**, 1—26 (1960).

Barrow, E. M., W. R. Bullock, and J. B. Graham: The carrier state in PTC deficiency. J. Lab. clin. Med. **55**, 936—946 (1960).

Beutler, E., M. Yeh, and V. F. Fairbanks: The normal human female as a mosaic of X-chromosome activity: studies using the gene for G-6-PD deficiency as a marker. Proc. nat. Acad. Sci. (Wash.) **48**, 9—16 (1962).

Davidson, R. G., H. M. Nitowsky, and B. Childs: Demonstration of two populations of cells in the human female heterozygous for glucose-6-phosphate dehydrogenase variants. Proc. nat. Acad. Sci. **50**, 481—485 (1963).

Gans, M.: Etude génétique et physiologique du mutant z de Drosophila melanogaster. Bull. Biol. France Belg. Suppl. **38**, 1—90 (1953).

Isselbacher, K. J., E. P. Anderson, K. Kurahashi, and H. M. Kalckar: Congenital galactosemia, a single enzymatic block in galactose metabolism. Science **123**, 635—636 (1956).

Kazazian, H. H. jr., W. J. Young, and B. Childs: X-linked 6-phosphogluconate dehydrogenase in Drosophila: Subunit associations. Science **150**, 1601—1602 (1965).

Lyon, M. F.: Gene action in the X-chromosome of the mouse (Mus musculus L.). Nature **190**, 372—373 (1961).

Muller, H. J.: Evidence of the precision of genetic adaptation. Harvey Lectures Ser. **43**, 165—229 (1947—1948).

Nishimura, E. T., T. Y. Kobara, S. Takahara, H. B. Hamilton, and S. C. Madden: Immunologic evidence of catalase deficiency in human hereditary acatalasemia. Lab. Invest. **10**, 333—340 (1961).

Rapaport, S. I., M. J. Patch, and F. J. Moore: Antihemophilic globulin levels in carriers of hemophilia A. J. clin. Invest. **39**, 1619—1625 (1960).

Robinson, A.: The assay of galactokinase and galactose-1-phosphate uridyl transferase activity on human erythrocytes. J. exp. Med. **118**, 359—370 (1963).

Stern, C.: Über die additive Wirkung multipler Allele. Biol. Zbl. **49**, 241 —290 (1929).

Takahara, S., H. B. Hamilton, J. B. Neil, T. Y. Kobara, Y. Ogura, and E. T. Nishimura: Hypocatalasemia: a new genetic carrier state. J. clin. Invest. **39**, 610—619 (1960).

Tanaka, K. R., W. N. Valentine, and S. Miwa: Pyruvate kinase (PK) deficiency hereditary nonspherocytic hemolytic anemia. Blood **19**, 267 —268 (1962).

Chapter 8

Various Consequences of the Dosage Compensation by X-inactivation

The invariably hemizygous expression of the X-linked gene in the mammalian female should result in the visibly mosaic appearance of heterozygous females if the female carries a mutant gene which affects individual melanocytes or epidermal cells.

Random inactivation of one or the other X early in embryonic life, on the other hand, should result in the sporadic occurrence of genetically heterozygous females which exhibit the uniformly hemizygous phenotype either for the wild-type allele or for the mutant allele.

The possibility that the dosage compensation by X-inactivation might not involve all the genes on the X shall also be considered in this chapter.

a) Visibly Mosaic Phenotypes Exhibited by Mammalian Females Heterozygous for X-linked Genes Affecting Hairs and Skin

Lyon and other mouse geneticists have been aware of the curious fact that most of the known X-linked mutant genes of the mouse are neither dominant nor recessive. In fact, Lyon was apparently inspired by this curious fact in formulating the hypothesis of dosage compensation by random inactivation of the individual X in the female.

As to the X-linked mutant genes which affect the formation of melanin granules by individual melanocytes, the six mutant genes of the mouse which dilute coat color give a mosaic phenotype to the heterozygous female; white or light-colored patches are intermingled with dark wild-type fur. These mutant genes are *Mottled* and *Brindled* (Fraser et al., 1953), *Tortoise-shell* (Dickie, 1954), *Dappled* (Phillips, 1961), *Dappled-2* (Lyon, 1960), and *Blotchy* (Russell and Saylors, 1962) (see Table 1). Although the males hemizygous for these mutant genes die *in utero,* the interpretation that white or light-colored patches of the heterozygous females are populated by these melanocytes which have the active mutant allele can be inferred from the fact that the affected hemizygous male fetuses, when examined, appear uniformly whitish.

The two X-linked mutant coat color genes of the golden hamster (*Mesocricetus auratus*) also manifest a mosaic phenotype in the heterozygous females. The male hemizygous for *Mottled-white* is thought to be entirely whitish, but he invariably dies *in utero*. The heterozygous females are mottled with whitish-gray patches in the same manner as the female mice heterozygous for mutant genes of the *Mottled* series (MAGALHAES, 1954). In the case of the other mutant gene *Tortoise-shell,* the hemizygous male is viable and entirely yellow, while patches of yellow are seen intermingled with wild-type fur in the heterozygous female (TAY, 1964). This latter mutation of the golden hamster may well be homozygous with the X-linked *Yellow* of the cat (ROBINSON, 1959).

In the cat, the X-linked *Yellow* is apparently allelic to the gene *Black*. When the black tomcat is mated to the yellow female, all the male kittens are yellow and all the female kittens are tortoise-shell, exhibiting vertical stripes of yellow hairs intermingled with vertical stripes of black hairs. When accompanied by the autosomal spotting gene, the phenotype of the heterozygous female changes to a prettier calico which demonstrates white extremities and large, distinct yellow and black patches on the dorsal region. The normal male cat carrying only one X should not manifest the tortoise-shell or calico phenotype. In Japan, it has been known for centuries that the males are occasionally born with the calico phenotype and are sterile. Recently, THULINE and NORBY (1961) showed that at least some of these calico males are the XXY, the feline counterpart of the human Klinefelter's syndrome.

The X-linked mutant genes which affect individual epidermal cells and cells of hair follicles are also expected to show a mosaic phenotype in heterozygous females. Three such genes are known in the mouse. The gene action of *Tabby* (FALCONER, 1953) is described in detail in Chapter 4, and the appearance of a hemizygous male is illustrated in Figure 12. The other two are *Striated* (PHILLIPS, 1963) and *Greasy* (RUSSEL and LARSEN, 1964). In each of these three genes, the heterozygous females demonstrate a mosaic phenotype having stripes of affected mutant areas in the background of the normal wild-type area.

The *streaked hairlessness* of cattle (ELDRIDGE and ATKESSON, 1953) is also apparently due to the regional expression of the mutant gene in the heterozygous female. The affected hemizygous male which invariably dies *in utero* is expected to be entirely hairless.

b) Invisible Mosaic Phenotypes of Heterozygous Females Revealed by the Microscope and Other Tools

In the case of many X-linked genes, the mosaic phenotypes exhibited by heterozygous females are not visible to the naked eye, but the coexistence of two cell populations in the target tissue of a particular gene locus can be revealed by more sophisticated means. The experiments carried out by BEUTLER and his colleagues (1962) and DAVIDSON and his colleagues (1963) on erythrocytes and fibroblasts of women having two different G-6-PD alleles are good examples. These experiments are described in Chapter 7.

In demonstrating the presence of the X-linked histocompatibility antigen in the mouse, BAILEY (1963) obtained hybrids between the two appropriate inbred strains. When skin of the F_1-female was grafted to her brother, skin was rejected in patches. The interpretation was that the areas of the female skin which were rejected were made of cells which had the paternally derived X in the functional state. As both parental lines were highly inbred, only the paternally derived X of the F_1-female genomes was foreign to the F_1-male. The other areas of the skin where the paternally derived X-chromosome was rendered inactive, were accepted by the F_1-female.

As shown in Table 2, pseudohypertrophic progressive muscular dystrophy of Duchenne type is one of the X-linked defective traits of man. The affected male usually dies before the age 20. The heterozygous female is apparently healthy most of the time and no visible mosaic effect can be seen. When muscular tissue of the heterozygous female was examined under the microscope howewer, it was found that affected pseudohypyertrophic muscle fibres were intermingled with normal muscle fibres (PEARSON et al., 1963).

FUDENBURG and HIRSCHHORN (1964) found that cultured lymphocytes from the male affected with X-linked agammaglobulinemia were incapable of differentiating into plasma cells; consequently, there was no γ-globulin synthesis. Cultured lymphocytes of the heterozygous female, on the other hand, were of two types: those which differentiated into plasma cells and synthesized γ-globulin, and others which failed to differentiate and were incapable of γ-globulin production.

c) Occasional Manifestation by Heterozygous Females of Entirely Hemizygous Phenotypes

In man and the cat, the fact that the sex chromatin body is first found in the late blastula stage of embryonic development (PARKS, 1957; AUSTIN and AMOROSO, 1957), suggests that random inactivation of one or the other X of the female occurs when there are only two to three thousand cells comprising the entire embryo (excluding extraembryonic and trophoblastic elements). At the time of X-inactivation, a very small number of progenitor cells must represent each entire adult tissue of the future. It is then expected that random inactivation occasionally leads to the situation where one whole tissue is populated exclusively by one or the other of the two cell types. Indeed, a certain proportion of the females heterozygous for certain X-linked genes do manifest the entirely hemizygous phenotype.

In the case of women who are heterozygous for the defective mutant G-6-PD allele, BEUTLER and his colleagues (1962) reasoned that if only two cells at the time of random inactivation form the basis of the entire erythroid series, one would expect to see the customary 50% reduction in the enzyme level only in one-half of the women. One-quarter should show the 100% enzyme level of the normal woman, while the other one-quarter should show zero enzyme level of the affected hemizygous male. There is, of course, no reason to believe that the number of erythrocyte progenitors is as small as two at the time of X-inactivation. Furthermore, the number of progenitors may vary from one embryo to the next. Thus, in reality, a great majority of heterozygous women demonstrate the 50% enzyme level expected of them. Nevertheless, it is a fact that an appreciable proportion of genetic heterozygotes for G-6-PD deficiency show either a normal or grossly deficient erythrocyte enzyme level (TRU-JILLO et al., 1961).

We do not know what cell type produces antihemophilic globulin (AHG), but RAPAPORT and his colleagues (1960) estimated that 2% of females heterozygous for the gene for hemophilia A have AHG levels sufficiently low to result in clinical manifestation of hemophilia. Conversely, a few per cent of heterozygous women demonstrate near normal AHG level. Similarly, FROTA-PESSOA and his colleagues (1963) have shown that a certain percentage of women heterozygous for hemophilia B demonstrate plasma thromboplastic component (PTC) levels either noticeably above or below the expected 50% level.

d) The Question of Exceptional X-linked Genes which Escape Inactivation in the Female

In biology we are so accustomed to finding exceptions to the rule that it was not surprising when suggestions were made that not all the X-linked genes of man, the mouse, and other placental mammals behave in conformity with the X-inactivation hypothesis proposed by Lyon.

If a segment of the X consistently were to escape heterochromatinization, individual somatic cells of the female would be endowed with two euchromatic doses of that segment, both alleles would function in synchronous harmony. Thus, the mammalian female would not be a natural mosaic with regard to the activity of X-linked genes. Consequently, these genes not being included in the dosage compensation mechanism should show a definite dosage effect. The direct product of such a gene produced by the male should be 50% of that produced by the female. Since it is against the economy of nature, it is unlikely that placental mammals evolved concurrently two different means of dosage compensation: one by random inactivation of one or the other X in the female for some of the X-linked genes, and the other by having modifier genes for the remainder of the X-linked genes.

Among the known X-linked genes of man (Table 2), the X-linked blood group gene locus (Xg) has often been cited as an example of exceptional genes which escapes inactivation in the female. When available evidence is closely scrutinized, it becomes apparent that this gene locus is not exceptional at all. The erythrocytes of males carrying the Xg^a (+ve) allele are agglutinated by suitable antisera, whereas those of males with the Xg (−ve) allele are not. Mann and his colleagues (1962) have found that the Xg^aY male individual reacts more strongly with antisera than the heterozygous Xg^aXg female and that Xg^aXg^a homozygotes react as strongly as hemizygotes and more strongly than heterozygotes. The dosage is clearly compensated for the Xg locus. However, several attempts to demonstrate the coexistence of positive and negative erythrocytes in the heterozygous female had failed (Gorman et al., 1963; Reed et al., 1963). The experiment of

REED and his colleagues undoubtedly failed on technical grounds (COHEN et al., 1964).

The strong indication that dosage of the Xg^a gene is indeed compensated by X-inactivation is given by the family studies. There have been three instances in which phenotypically Xg (−ve) mothers gave birth to Xg^a (+ve) sons (SANGER et al., 1964; CHOWN et al., 1964). The genotype of these three mothers have to be Xg^aXg; but by chance, the entire erythropoietic system of each is derived from one type of progenitor cell which inactivated the X carrying the Xg^a (+ve) allele.

Recently, a direct evidence that Xg^aXg heterozygotes indeed have two populations of erythrocytes has been obtained by MACDIARMID and his colleagues (1967). One of the two X-chromosomes of a female child carried both the Xg (+ve) allele and the gene for Cooly-Rundles-Falls type hypochromic anemia (see Chapter 4 and Table 2), while the other X carried the Xg (−ve) allele and the wild-type allele of hypochromic anemia. She was in fact a double heterozygote. When her erythrocytes were separated by differential centrifugation in layered gum acacia solutions, normal erythrocytes were Xg (−ve), while abnormal hypochromic microcytes were Xg (+ve). Her mother who was also heterozygous for hypochromic anemia was homozygous for Xg^a. In her case, both normal and microcytic erythrocytes were Xg (+ve).

In the mouse, two genes, *scurfy* (RUSSELL et al., 1959) and *Sparse-fur* (RUSSELL, 1964), both affecting coat texture, have been cited as exceptional genes. The females heterozygous for these two genes do not exhibit mosaic phenotypes. However, no evidence has been presented which reveals the dosage effect of these two genes. As shall be shown in Chapter 12, many Z-linked genes of avian species are not compensated for dosage. In each of these dominant traits, only the homozygous male shows the full manifestation of a mutant phenotype, while the hemizygous female is a look-alike of the heterozygous male. *Scurfy* is apparently a recessive trait which suggests that an amorphic allele might be responsible. Even in absence of the dosage compensation, an amorphic allele is not expected to show the dosage effect. Thus, no judgement can be made as to whether or not this gene locus on the X-chromosome of the mouse escapes the dosage compensation by X-inactivation. *Sparse-fur,* on the other hand, is a dominant trait, yet it is lethal in the hemizygous state. This seems to suggest that this gene locus is involved in the dosage compensation mechanism. If not, the hemizygous male should be as well off as the heterozygous female, and only the homozygous female should die *in utero*. The apparent absence of mosaicism in the females heterozygous for *scurfy* as well as *Sparse-fur* is probably due to the fact that the observed effects of these two genes are not produced by localized action in individual epidermal cells, but by some non-localized systemic effect.

Taking advantage of SEARLE's X-autosome translocation, LYON (1966) revealed that the gene locus for *scurfy* is indeed subjected to inactivation in the female. As described in Chapter 10 and illustrated in Figure 22, preferential inactivation of the normal X occurs in female mice which are heterozygous for SEARLE's translocation. When the *scurfy* gene was put on the translocation-bearing X, and the wild-type allele of *scurfy* on the normal X, such a female demonstrated full manifestation of the *scurfy* phenotype.

It then appears that today we have no good example of exceptional X-linked genes which are not involved in the dosage compensation mechanism by random inactivation of one or the other X in the female. Nevertheless, it is quite conceivable that some of the X-linked genes might be relatively insensitive to the inactivating influence of heterochromatinization. After all, the entirely heterochromatic Y-chromosome of *Drosophila* carries the functional gene *bobbed* (Chapter 7). These X-linked genes, if they exist, would be expected to show both some dosage effect and semi-hemizygous expression in individual somatic cells of the heterozygous female.

e) Hemizygous Expressions of Certain Autosomal Genes in Mammals

In the case of X-linked genes, an invariably hemizygous expression in individual somatic cells of the female is a means to nullify the genetic disparity that exists between the male with only one X and the female with two X's. The first event which occurred during the course of evolution must have been the doubling of the rate of product output of each X-linked gene. In this way, the X-linked genes were able to accommodate themselves to the hemizygous existence in the male. The dosage compensation by inactivation of one of the two X's of the female was a natural next step.

Unlike X-linked genes, autosomal genes always exist in double doses in both sexes. There has been no need for the autosomal genes to double their efficiency. The monosomic condition for even a small autosome is apparently lethal in placental mammals. This is a good indication that doubling of the rate of product output has not occurred to autosomal genes during evolution. For many autosomal genes, direct evidence actually exists which reveals that both alleles function simultaneously in each diploid somatic cell. Persons who are heterozygous for the sickle cell anemia have two different alleles at the autosomal gene locus for the β-polypeptide chain of hemoglobin. Both normal hemoglobin $(a_2{}^A\beta_2{}^A)$ and abnormal hemoglobin S $(a_2{}^A\beta_2{}^S)$ are found in each erythrocyte of such a heterozygote, proving that both the normal β^A-allele and abnormal β^S-allele are functioning simultaneously. The formation of the hybrid dimer enzyme by individuals heterozygous for the electrophoretic variants of certain autosomally inherited enzymes such as 6-PGD, as mentioned in Chapter 7, is another such direct evidence.

In view of the above, the curious fact about certain autosomal genes of placental mammals should be mentioned here. Phenotypic expressions of these autosomal genes are invariably hemizygous (homozygous?). Consequently, heterozygotes are recognizable mosaics.

In the mouse, the coat color mutant gene *Varitint-waddler (Va)* is located on the linkage group XVI autosome. An animal homozygous for this mutation is uniformly white, but the phenotype of the heterozygote is regularly a mosaic of wild-type, white and light colored patches. Thus, *Va*-heterozygotes appear very similar to the female mice heterozygous for one of the X-linked mutant genes of the *Mottled* series (CLOUDMAN and BUNKER, 1945). It appears that parts of the coat of *Va*-heterozygotes are populated by cells with the active *Va*-allele, while in the remaining parts of the coat, the wild-type allele is functioning. It is unfortunate that the linkage group XVI is an unexplored linkage group having no other pertinent marker genes.

More striking examples of invariably hemizygous (homozygous?) expressions of autosomal genes are provided by the structural genes for immuno-globulin molecules. As stated in previous chapters, the structural genes for light (L)- and heavy (H)-polypeptide chains of immuno-globulin molecules are located on the autosomes of man, the rabbit, and the mouse. These two gene loci are either on two different autosomes or on two ends of the same autosome. In these species, allelic polymorphism of L- and H-chain gene loci is known as allotypy. In man, allelic polymorphism at the L-chain gene locus is known as the Inv-system, and that of the H-chain gene locus as Gm-system. Gamma-1-G-globulin of multiple myeloma in a patient genetically heterozygous at the H-chain gene locus has been shown to be the product of one or the other allele (HARBOE et al., 1962; KUNKEL et al., 1964). If each myeloma originates from a single cell, this indicates that man normally has two populations of immuno-globulin-producing cells with regard to the activity of H-chain alleles. This peculiarity of immuno-globulin production was further clarified by PERNIS and his colleagues (1965). In the rabbit *(Oryctolagus cuniculus),* the L-chain gene locus is known as Ab-locus, while the so-called Aa-locus represents the H-chain gene locus. In animals heterozygous at either the Aa- or Ab-locus as well as in rare double heterozygotes, it was found that while both alleles of each of the two loci were apparently functioning in individual "blast" cells of the germinal centers of spleen and lymph nodes, each plasma cell of medullary cords was producing only one type of doubly homozygous immuno-globulin molecules. When histological sections of spleen and lymph nodes of a double heterozygote (Aa1, Aa2, Ab4, Ab6) were stained with fluorescent anti-Aa1 and anti-Aa2 sera in succession, "blast" cells were double stained, while individual plasma cells were stained either by anti-Aa1 or anti-Aa2, but never by both. The same occurred when anti-Ab4 and anti-Ab6 sera were used in succession. Thus, a doubly heterozygous rabbit has four populations of plasma cells with regard to the activity of autosomal genes which produce subunit polypeptides of immuno-globulin molecules. The phenotypes of four cell populations are: (1) Aa1, Ab4; (2) Aa1, Ab6; (3) Aa2, Ab4; and (4) Aa2, Ab6. Such hemizygous expressions of the autosomal genes in each individual plasma cell may be essential to the antibody production mechanism.

Two antibodies of distinct specificity differ from each other by a contiguous, variable amino acid sequence of both the L-chain and H-chain. Whether this variable amino acid sequence was generated during evolution by gene duplication and mutation or during differentiation by a higher mutation mechanism, the odds of the same change in base sequence occurring simultaneously in both alleles at the L-chain gene locus of a single diploid cell is phenomenally small. The same can be said of the two alleles at the H-chain locus. Thus, only by hemizygous expression of both the L- and H-chain genes can an individual plasma cell and its descendants become fully committed to the production of a unique antibody.

Today, the coat-color mutant gene *Varitint-waddler* of the mouse and the L- and H-chain genes for γ-globulin are the only autosomal genes of mammals that we are aware of which demonstrate hemizygous expression, and therefore manifest the mosaic effect in heterozygotes. As we learn more about direct products of individual autosomal genes, it may turn out that a considerable number of the autosomal genes actually behave like the X-linked genes in females in that only one or the other allele expresses itself in each individual somatic cell.

It may be that the mechanism of random inactivation of one or the other allele early in embryonic life which is utilized by the X-linked genes as a means of achieving the dosage compensation is also utilized by some of the autosomal genes. Conversely, the underlying mechanism which causes the hemizygous (homozygous?) expression in individual somatic cells of the autosomal genes might be quite different from that which operates on the X-linked genes of the female. Homozygous expression of autosomal genes by individual somatic cells can be the result of somatic segregation which preferentially involves certain autosomal linkage groups. Two homozygous cell types can be reconstituted within an individual heterozygous for autosomal genes by means of tetrapolar mitosis of a tetraploid cell or of predirected double non-disjunction involving both members of a homologous pair (OHNO et al., 1966).

References

AUSTIN, C. R., and E. C. AMOROSO: Sex chromatin in early cat embryos. Exp. Cell Res. **13**, 419—421 (1957).

BAILEY, D. W.: Mosaic histocompatibility of skin grafts from female mice. Science **141**, 631—634 (1963).

BEUTLER, E., M. YEH, and V. F. FAIRBANKS: The normal human female as a mosaic of X-chromosome activity: studies using the gene for G-6-PD deficiency as a marker. Proc. nat. Acad. Sci. (Wash.) **48**, 9—16 (1962).

CHOWN, B., M. LEWIS, and H. KAITA: The Xg blood group system: Data on 294 white families, mainly Canadian. Canad. J. Genet. Cytol. **6**, 431—434 (1964).

CLOUDMAN, A. M., and L. E. BUNKER: The varitint-waddler mouse. J. Hered. **36**, 259—263 (1945).

COHEN, F., W. W. ZUELZER, and M. M. EVANS: Fluorescent-antibody technique and the Lyon hypothesis. Lancet **I**, 1392—1393 (1964).

Davidson, R. G., H. M. Nitowsky, and B. Childs: Demonstration of two populations of cells in the human female heterozygous for glucose-6-phosphate dehydrogenase variants. Proc. nat. Acad. Sci. **50**, 481—485 (1963).

Dickie, M. M.: The tortoise-shellhouse mouse. J. Hered. **145**, 158—190 (1954).

Eldridge, F., and F. W. Atkeson: Streaked hairlessness in Holstein-Friesian cattle: A sex-linked lethal character. J. Hered. **44**, 265—271 (1953).

Falconer, D. S.: Total sex-linkage in the house mouse. Z. Indukt. Abstamm.- u. Vererb.-L. **85**, 210—219 (1953).

Fraser, A. S., S. Sobey, and C. C. Spicer: *Mottled,* a sex-modified lethal in the house mouse. J. Genet. **51**, 217—221 (1953).

Frota-Pessoa, O., E. L. Gomes, and T. R. Calicchio: Christmas factor: dosage compensation and the production of blood coagulation factor IX. Science **139**, 348—349 (1963).

Fudenberg, H. H., and K. Hirschhorn: Agammaglobulinemia: the fundamental defect. Science **145**, 611—612 (1964).

Gorman, J. G., J. Di Re, A. M. Treacy, and A. Cahan: The application of Xg^a antiserum to the question of red cell mosaicism in female heterozygotes. J. Lab. clin. Med. **61**, 642—649 (1963).

Harboe, M., C. K. Osterland, M. Mannik, and H. G. Kunkel: Genetic characters of human gamma globulins in myeloma proteins. J. Exp. Med. **116**, 719 (1962).

Kunkel, H. G., J. C. Allen, H. M. Grey, L. Martensson, and R. Grubb: A relationship between the H-chain groups of 7S gamma globulin and the Gm system. Nature **203**, 413 (1964).

Lyon, M. F.: A further mutation of the *mottled* type in the house mouse. J. Hered. **51**, 116—121 (1960).

— Lack of evidence that inactivation of the mouse X-chromosome is incomplete. Genet. Res. Camb. **8**, 197—203 (1966).

MacDiarmid, W. D., G. R. Lee, G. E. Cartwright, and M. M. Wintrobe: X-inactivation in an unusual X-linked anemia and the Xg^a group. Clinic. Res. **15**, 132 (1967).

Magalhaes, H.: *Mottled-white,* a sex-linked lethal mutation in the golden hamster, *Mesocricetus auratus.* Anat. Rec. **120**, 752 (1954).

Mann, J. D., A. Cahan, A. G. Gelb, N. Fisher, J. Hamper, P. Tippett, R. Sanger, and R. R. Race: A sex-linked blood group. Lancet I, 8—10 (1962).

Ohno, S., C. Weiler, J. Poole, L. Christian, and C. Stenius: Autosomal polymorphism due to pericentric inversions in the deer mouse *(Peromyscus maniculatus)* and some evidence of somatic segregation. Chromosoma **18**, 177—187 (1966).

Parks, W. W.: The occurrence of sex chromatin in early human and macaque embryos. J. Anat. (Lond.) **91**, 369—373 (1957).

Pearson, C. M., W. M. Fowler, and S. W. Wright: X-chromosome mosaicism in females with muscular dystrophy. Proc. nat. Acad. Sci. **50**, 24—31 (1963).

PERNIS, B., G. CHIAPPINO, A. S. KELUS, and P. G. H. GELL: Cellular localization of immunoglobulins with different allotypic specificities in rabbit lymphoid tissues. J. exp. Med. **122**, 853—876 (1965).

PHILLIPS, R. J. S.: *"Dappled"*, a new allele at the *Mottled* locus in the house mouse. Genet. Res. **2**, 290—295 (1961).

— *Striated*, a new sex-linked gene in the house mouse. Genet. Res. **4**, 151—153 (1963).

RAPAPORT, S. I., M. J. PATCH, and F. J. MOORE: Antihemophilic globulin levels in carriers of hemophilia A. J. clin. Invest. **39**, 1619—1625 (1960).

REED, T. E., N. E. SIMPSON, and B. CHOWN: The Lyon Hypothesis. Lancet **II**, 467—468 (1963).

ROBINSON, R.: Genetics of the domestic cat. Bibl. genet. **18**, 273—362 (1959).

RUSSELL, W. L., L. B. RUSSELL, and J. S. GOWER: Exceptional inheritance of a sex-linked gene in the mouse explained on the basis that the XO-constitution is female. Proc. nat. Acad. Sci. (U.S.) **45**, 554—560 (1959).

RUSSEELL, L. B., and C. L. SAYLORS: Induction of paternal sex-chromosome losses by irradiation of mouse spermatozoa. Genetics **47**, 7—10 (1962).

— Another look at the single-active-X hypothesis. Trans. N.Y. Acad. Sci. Ser. II **26**, 726—736 (1964).

—, and M. M. LARSEN: Personal communication. Mouse news letter **30**, 47 (1964).

SANGER, R., R. R. RACE, P. TIPPETT, J. GAVIN, R. M. HARDISTY, and V. DUBOWITZ: Unexplained inheritance of the Xg groups in two families. Lancet **I**, 955—956 (1964).

TAY, B.: A new line in hamsters — but strictly female. Daily Mirror, Aug. 22, P. 13 (1964).

THULINE, H. C., and D. E. NORBY: Spontaneous occurrence of chromosome abnormality in cats. Science **134**, 554—555 (1961).

TRUJILLO, J., V. FAIRBANKS, S. OHNO, and E. BEUTLER: Chromosomal constitution in glucose-6-phosphate dehydrogenase deficiency. Lancet **II**, 1454—1455 (1961).

Chapter 9

The Conservation of the Original X and Dosage Compensation in the Face of X-polysomy

From the variety of X-polysomic conditions described in man in recent years, one remarkable rule emerged: the number of sex chromatin bodies always equals the number of X-chromosomes present minus one. The dosage compensation mechanism which nullifies the genetic disparity between the male and the female also serves to minimize the detrimental consequences of X-polysomy. But when this rule is viewed against the broad background of knowledge of the whole infraclass *Eutheria,* it becomes just a small portion of the grand scheme of dosage compensation. In this chapter, it will be shown that the dosage compensation mechanism is universal to all the placental mammals in a qualitative as well as a quantitative sense in that the aim is to preserve only one original X in the functional state in each diploid somatic cell under all the normal and abnormal conditions.

a) X-chromosome Behavior in Individuals with Various Abnormal X-chromosome Constitutions

Until 1959 it was a concensus that in such highly evolved organisms as man, any aneuploidy which reflects gross imbalance of genetic material would surely be lethal. In 1959, however, it was shown by LEJEUNE and his colleagues (1959) that individuals afflicted with the syndrome of mongolian idiocy are trisomic for one of the smallest autosomes. Simultaneously, the sterile female of short stature with the Turner syndrome was shown to have the XO-constitution (FORD et al., 1959), and the sterile male with the Klinefelter syndrome, the XXY-constitution (JACOBS and STRONG, 1959). Since then, a truly astounding variety of abnormal chromosome constitutions have been found in abnormal, but viable, individuals of man. Even though the trisomic condition for even very small autosomes are severely detrimental (PATAU et al., 1960; EDWARDS et al., 1960), and the autosomal monosomy is barely compatible with viability only if a small segment of a chromosome is involved (DE GROUCHY et al., 1964), individuals with various abnormal sex chromosome constitutions do quite well, aside from the fact that they are usually sterile and tend to be mentally retarded. In addition to the XO- and the XXY-constitutions already mentioned, the following types of X-polysomies have been found: the XXX-female (JACOBS et al., 1959), the XXXY-male (FERGUSON-SMITH et al., 1960), the XXXX-female (CARR and BARR, 1961), and the XXXXY-male (MILLER et al., 1961). The indications are that the X-chromosome carries as much of a load of Mendelian genes as an autosome of comparable size. Thus, the automatic inactivation of extra X-chromosomes emerges as the factor responsible for relative harmlessness of the X-polysomy. When there are three X-chromosomes in the complement (2AXXX or 2AXXXY), two of the three X's become heterochromatic and therefore inactivated. Thus, two sex chromatin bodies are regularly seen in each somatic nucleus of such individuals. Similarly, the XXXX-female and XXXXY-male individuals demonstrate three sex chromatin bodies per nucleus; three of the four X-chromosomes are inactivated. Whatever the number of extra X-chromosomes added to the normal diploid complement, the dosage compensation mechanism preserves only one functional X in each diploid somatic cell. The direct genetic evidence in support of inactivation of all the X's except one, has been obtained by GRUMBACH and his colleagues (1962). They studied erythrocyte glucose-6-phosphate dehydrogenase levels of 15 individuals, including 9 with three or four X-chromosomes. All had normal enzyme levels except for one XXXX-female whose value was slightly higher.

b) Pattern of X-inactivation in Rodent Species with Exceptionally Large X-chromosomes

The heterochromatinization of one entire X of the female mentioned thus far occurs in a great majority of placental mammals including man, the mouse, the dog, the horse, and cattle. The X-chromosome of these species comprises about 5% of the genome. These species apparently conserved the original X-chromosome of a common ancestor in its entirety. Among rodents, however, there is a number of exceptional species that are endowed with X-chromosomes of a much larger size. These larger X's may have been produced by literal duplication, triplication, and quadruplication of the original X during the course of evolution. In these, in addition to one entire X of the female, heterochromatinization extends to an excessive part of the euchromatic X of the male as well as the female. The studies on these exceptional rodent species revealed that the universal aim of the dosage compensation mechanism in placental mammals is indeed in the conservation of one genetic equivalent of the original X in each somatic cell.

The X-chromosome comprising 10% of the genome was termed the duplicate-type X (OHNO et al., 1964). The duplicate-type X was found in Eurasian hamsters, such as the Chinese hamster *(Cricetulus griseus*, 2n = 22), the golden hamster *(Mesocricetus auratus*, 2n = 44), and the European hamster *(Cricetus cricetus*, 2n = 22), as well as in the rat-like rodent of Africa *(Mastomys coucha*, 2n = 32) (HUANG and STRONG, 1962) and the chinchilla *(Chinchilla laniger*, 2n = 64). Excluding the chinchilla, the duplicate-type X of these species mentioned above was accompanied by the proportionately large Y-chromosome. In the chinchilla, however, the Y remained a minute element.

It is of interest to note that the very first species of placental mammals studied by the radioautography using tritiated thymidine was the Chinese hamster. The late DNA replication of the heterochromatic X in the female was first found on the exceptional duplicate-type X of this species (TAYLOR, 1960). The X of the Chinese hamster is the submediocentric element. The subterminal Y is almost as large as the X (YERGANIAN, 1959). In the male somatic cells, TAYLOR found that the entire Y and a great part of the long arm of the X which measured 50% of the total length synthesized their DNA at the very end of the synthetic phase of interphase. Of the two X-chromosomes of the females, one showed exactly the same labeling pattern as the male X, while the other X was entirely late-labeling. Thus, somatic cells of both sexes had only one-half of one X in the euchromatic state, and only this euchromatic segment synthesized DNA in synchrony with the autosomes. This pioneering discovery on late replication of the heterochromatic segments of the sex chromosomes of the Chinese hamster is in sharp contrast to the subsequent findings on labeling patterns of the original-type X of man (GILBERT et al., 1962; GERMAN, 1962), the mouse (GALTON and HOLT, 1965), and other placental mammals. In these species with the original-type X, one whole X remains in an euchromatic state in both the male and the female. It is rather strange that the significance of this fundamental difference in labeling patterns was not recognized until 1964 (OHNO et al., 1964).

The duplicate-type X of the golden hamster is the mediocentric element, and the subterminal Y which is entirely heterochromatic is again almost as large as the X. In addition to one entire X of the female, one arm of the other X of the female as well as one arm of the male X manifest the heterochromatic condition (SAKSELA and MOORHEAD, 1962) and show the asynchronous late DNA replication (GALTON and HOLT, 1964). Similarly, only one arm of one X remains euchromatic in somatic cells of both sexes of the European hamster.

8*

The duplicate-type X of the European hamster is also a mediocentric element (WOLF and HEPP, 1966).

Among the species with the duplicate-type X, the chinchilla was exceptional in two aspects: (1) the increase in size of the X was not accompanied by a proportionate increase in size of the Y (the Y remained a minute element); (2) the functional X of both sexes was divided into three segments instead of two. The entire short arm and the distal end of the long arm of the submediocentric X comprised the late replicating segments, and the euchromatic segment was inserted between the two. Nevertheless, of the two X-chromosomes of female somatic cells, one was entirely late labeling with tritiated thymidine while the other X showed the euchromatic segment comprising about 50% of the total length in the middle. The male X also demonstrated the two late replicating segments on both sides of the euchromatic segment. The minute Y was entirely late replicating (GALTON et al., 1965).

Because inactivation by heterochromatinization extends to one-half of the functional X, a sexual dimorphism of somatic interphase nuclei is not distinct in these species with the duplicate-type X. Male interphase nuclei are not sex-chromatin-negative, but positive or even double positive. The chromocenter made by one-half of the duplicate-type X is naturally of the same size as the female-specific chromocenter of man and other species which are made of one whole original-type X. Except for the chinchilla, which has a minute Y, there should be another prominent chromocenter in male nuclei which represents the very large Y in its entirety. Female interphase nuclei should also

---

Fig. 20. Photomicrographs which reveal the pattern of inactivation of the quadruplicate-type X of the field vole *(Microtus agrestis)*. From the paper by WOLF and his colleagues (WOLF et al., 1965). 1) Male somatic interphase nuclei, each with two enormous chromocenters. 2) and 3) A male mitotic metaphase figure which has incorporated tritiated thymidine into its DNA near the end of the synthetic phase of a previous interphase. Comparing the photograph taken before the autoradiography (2) with that taken afterward (3), it may be noted that the entire Y (8 O'clock) as well as three-fourths of the X (5 O'clock) are late labeling, and therefore heterochromatic. 4) Female somatic interphase nuclei also displaying two prominent chromocenters. 5) and 6) A female somatic metaphase figure with one entirely late labeling X (3 O'clock), and the other X showing late labeling of three-quarters of its length (2 O'clock)

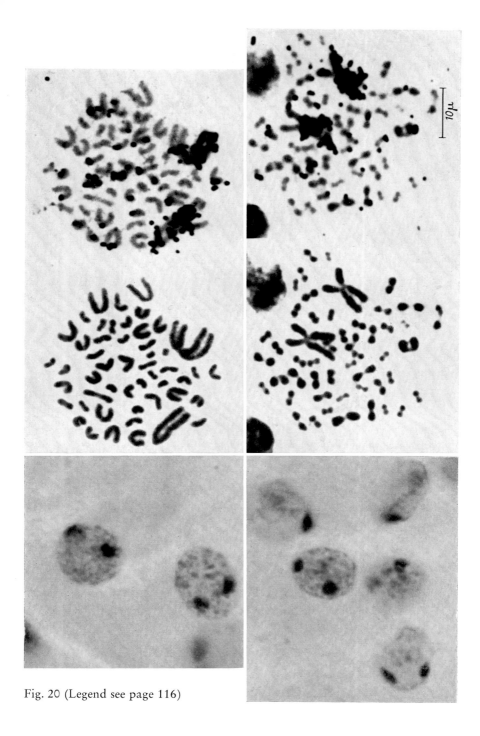

Fig. 20 (Legend see page 116)

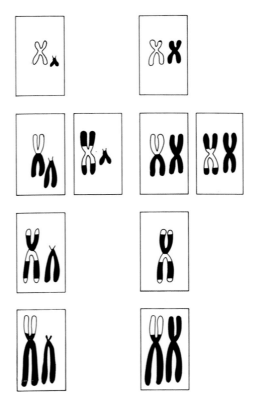

Fig. 21. Schematic representation of the behavior pattern of various types of X-chromosomes in male and female somatic cells of mammals; the males are represented on the left and the females, on the right. Euchromatic parts of the X are outlined, while the heterochromatic parts of the X as well as the Y are painted solid black. Top row: The original-type X, comprising 5% of the genome, is found in a great majority of mammals including man, the mouse and the cat. In the male (left), only the minute Y is heterochromatic while in the female, one entire X is heterochromatic. 2nd row: The duplicate-type X found in Eurasian hamsters, the mastomys and the chinchilla. In the male, one-half of the X and the entire Y are heterochromatic. In the female, one entire X and one-half of the other X are heterochromatic. Because the chinchilla is exceptional in two ascepts (see text), the pattern of inactivation of the X of this species is shown separately at the right of the male as well as the female column. 3rd row: The triplicate-type X found in the creeping vole *(Microtus oregoni)*. In the male, two-thirds of the X as well as the entire Y are heterochromatic. The female of this species is normally XO, yet two-thirds of a single X is heterochromatic. Bottom row: The quadruplicate-type X of the field vole *(Microtus agrestis)*. In the male, three-fourths of the X and the entire Y are heterochromatic while in the female, one entire X as well as three-fourths of the other X are heterochromatic (see Fig. 20)

be sex-chromatin-double-positive; there should be one chromocenter made of one-half of one X and the other larger chromocenter made of one whole X.

The triplicate-type X comprising 15% of the genome has been found in the creeping vole *(Microtus oregoni, 2n = 17/18)* of the subfamily *Microtinae* which has a unique sex-determining mechanism. The sex-determining mechanism of this species shall be described in detail in Chapter 11. The female of this species normally starts from the XO-zygote, and somatic cells maintain the XO-constitution. Yet, each somatic interphase nucleus of the female demonstrates the presence of one very prominent chromocenter which is about twice the size of the female-specific sex chromatin body of other species with the original-type X. This is due to the fact that two-thirds of a single triplicate-type X in the female manifests the heterochromatic condition. Somatic cells of the male of this species maintain the ordinary XY-constitution, but the Y is almost as large as the triplicate-type X and is entirely heterochromatic. Accordingly, male interphase nuclei are sex-chromatin-double-positive. One chromocenter is made of two-thirds of one X and the other by an entire Y (OHNO et al., 1966).

The field vole *(Microtus agrestis, 2n = 50)* of Europe, also belonging to the subfamily *Microtinae*, probably has the largest X of all the placental mammals as it is the quadruplicate-type X comprising 20% of the genome. The Y is also enormous, making up 12% of the genome. SACHS and DANON (1956) have previously shown that somatic interphase nuclei of both sexes of this species demonstrate two very large chromocenters. WOLF and his colleagues (1965) revealed that of the two X-chromosomes of the female, one is entirely late replicating, and three-fourths of the other X is also late labeling. In the male, the entire Y and three-fourths of the X were late replicating (Fig. 20). Because of this extensive inactivation, the amount of euchromatic and functional X-chromosome material maintained by each somatic cell of both sexes of the field vole is still about the same as that maintained by each somatic cell of a great majority of placental mammals which are endowed with the original-type X. Figure 21 schematically illustrates the pattern of heterochromatinization in somatic cells of the three types of exceptionally large X-chromosomes described above.

c) Preferential Inactivation of an Abnormal X in the Female with one Normal X and one Abnormal X

Among sterile women with gonadal dysgenesis who are not XO, both a partially deleted X and an X-chromosome which is partially deleted and duplicated at the same time have been found. FRACCARO and his colleagues (1960) were the first to find the isochromosome for the long arm of the X. A woman with the Turner syndrome demonstrated the coexistence of two cell populations. One had the normal XX-complement, while in the other, one of the two X-chromosomes was replaced by a very large mediocentric chromosome. This abnormal chromosome was interpreted to have been made of two long arms and no short arm of the X. Later, a considerable number of women with the Turner syndrome were found to have the chromosome complement which includes one normal X and one isochromosome for the long arm of the X. When sex chromatin tests were performed on these women, the sex chromatin body was exceptionally large in every somatic interphase nucleus. This suggested that in this situation, the female was no longer a mosaic, but had only one population of somatic cells which had the normal X in the functional state and the abnormal isochromosome in the heterochromatic, inactive state. Indeed, when these cells were labeled by tritiated thymidine, it was found that the abnormal isochromosome was always a late replicating element (MULDAL et al., 1963).

A woman with one normal X and one deleted X which had lost the long arm (JACOBS et al., 1960), as well as a woman with one normal X and one deleted X which had lost the short arm (JACOBS et al., 1961), have been found. In both instances, there was preferential inactivation of the abnormal X. Interphase nuclei of these two women always contained the sex chromatin body of an unusually small size.

d) Brief Summary on Dosage Compensation in the Face of X-polysomy

Various findings described in this chapter reveal the universal aim of the dosage compensation mechanism of placental mammals. The aim is to preserve one functional unit of the original X in each somatic cell of both sexes under all circumstances.

In species with the original-type X such as man, no matter how many X-chromosomes are added to the complement, all X's except one whole X are automatically heterochromatinized and inactivated. Therefore, the number of sex chromatin bodies seen in individual nuclei are always the number of X-chromosomes present minus one. When one normal X is accompanied by an abnormal X, whether it be partially deleted or duplicated, random inactivation of one or the other X is no longer possible. In this situation, the female has no choice but to preferentially inactivate an abnormal X and preserve one whole normal X as a functioning unit.

In those exceptional species where duplication, triplication, and quadruplication of the original X took place during the course of evolution, heterochromatinization extends to the excessive part of the functional X of the male as well as the female. Again, only one functional segment equivalent to one whole original X is preserved in each somatic cell.

References

CARR, D. H., and M. L. BARR: An XXXX sex chromosome complex in two mentally defective females. Canad. med. Ass. J. **84**, 131—137 (1961).

DE GROUCHY, J., M. ARTHUIS, C. SALMON, M. LAMY et S. THIEFFRY: Le syndrome du cri du chat une nouvelle observation. Anal. Génétique **7**, 13—16 (1964).

EDWARDS, J. H., D. G. HARNDEN, A. H. CAMERON, M. CROSSE, and O. H. WOLFF: A new trisomic syndrome. Lancet I, 787 (1960).

FERGUSON-SMITH, M. A., A. W. JOHNSTON, and S. D. HANDMAKER: Primary amentia and micro-orchidism associated with an XXXY sex-chromosome constitution. Lancet **II**, 184—187 (1960).

FORD, C. E., D. W. JONES, P. E. POLANI, J. C. ALMEIDA, and J. H. BRIGGS: A sex chromosome anomaly in a case of gonadal dysgenesis (Turner's syndrome). Lancet I, 711—713 (1959).

FRACCARO, M., D. IKKOS, J. LINDSTEN, and K. KAIJSER: A new type of chromosomal abnormality in gonadal dysgenesis. Lancet **II**, 1144—1145 (1960).

GALTON, M., and S. F. HOLT: DNA replication patterns of the sex chromosomes in somatic cells of the Syrian hamster. Cytogenetics **3**, 97—111 (1964).

—, K. BENIRSCHKE, and S. OHNO: Sex chromosomes of the chinchilla: Allocycly and duplication sequence in somatic cells and behavior in meiosis. Chromosoma (Berl.) **16**, 668—680 (1965).

—, and S. F. HOLT: Asynchronous replication of the mouse sex chromosomes. Exp. Cell Res. **37**, 111—116 (1965).

GERMAN, J. L.: DNA synthesis in human chromosomes. Trans. N. Y. Acad. Sci. **24**, 395—407 (1962).

GILBERT, C. W., S. MULDAL, L. G. LAJTHA, and J. ROWLEY: Time-sequence of human chromosome duplication. Nature **195**, 869—873 (1962).

GRUMBACH, M. M., P. A. MARKS, and A. MORISHIMA: Erythrocyte glucose-6-phosphate dehydrogenase activity and X-chromosome polysomy. Lancet **I**, 1330—1331 (1962).

HUANG, C. C., and L. C. STRONG: Chromosomes of the African mouse. J. Hered. **53**, 95—99 (1962).

JACOBS, P. A., A. G. BAIKIE, W. M. COURT-BROWN, T. N. MACGREGOR, M. MACLEAN, and D. G. HARNDEN: Evidence for the existence of the human "super female". Lancet **II**, 423—425 (1959).

—, and J. A. STRONG: A case of human intersexuality having a possible XXY sex-determining mechanism. Nature **183**, 302 (1959).

—, D. G. HARNDEN, W. M. COURT-BROWN, J. GOLDSTEIN, H. G. CLOSE, T. N. MACGREGOR, N. MACLEAN, and J. A. STRONG: Abnormalities involving the X-chromosome in women. Lancet **I**, 1213—1216 (1960).

— —, K. E. BUCKTON, W. M. COURT-BROWN, M. J. KING, J. A. MCBRIDE, T. N. MACGREGOR, and N. MACLEAN: Cytogenetics studies in primary amenorrhoea. Lancet **I**, 1183—1188 (1961).

LEJEUNE, J., M. GAUTIER et R. TURPIN: Etude des chromosomes somatiques de neuf enfants mongoliens. C. R. Acad. Sci. **248**, 1721—1722 (1959).

MILLER, O. J., W. R. BREG, R. D. SCHMICKEL, and W. TRETTER: A family with an XXXXY male, a leukemic male and two 21-trisomic mongoloid females. Lancet **II**, 78—79 (1961).

MULDAL, S., C. W. GILBERT, L. G. LAJTHA, J. LINDSTEN, J. ROWLEY, and M. FRACCARO: Tritiated thymidine incorporation in an isochromosome for the long arm of the X chromosome in man. Lancet **I**, 861—863 (1963).

OHNO, S., W. BEÇAK, and M. L. BEÇAK: X-autosome ratio and the behavior pattern of individual X-chromosomes in placental mammals. Chromosoma **15**, 14—30 (1964).

—, C. STENIUS, and L. CHRISTIAN: The XO as the normal female of the creeping vole (*Microtus oregoni*). In: Chromosomes today, Vol. I, pp. 182—187. Eds. C. D. Darlington and K. R. Lewis. Edinburgh and London: Oliver and Boyd 1966.

PATAU, K., D. W. SMITH, E. THERMAN, S. L. INHORN, and H. P. WAGNER: Multiple congenital anomaly caused by an extra autosome. Lancet **I**, 790—793 (1960).

SACHS, L., and M. DANON: Diagnosis of the sex chromosomes in human tissues. Genetica **28**, 201—216 (1956).

SAKSELA, E., and P. S. MOORHEAD: Enhancement of secondary constrictions and the heterochromatic X in human cells. Cytogenetics **1**, 225—244 (1962).

TAYLOR, J. H.: Asynchronous duplication of chromosomes in cultured cells of Chinese hamster. J. Biophys. Biochem. Cytol. **7**, 455—464 (1960).

WOLF, U., G. FLINSPACH, R. BÖHM und S. OHNO: DNS-Reduplikationsmuster bei den Riesen-Geschlechtschromosomen von *Microtus agrestis*. Chromosoma **16**, 609—617 (1965).

Wolf, U., und D. Hepp: DNS-Reduplikationsmuster der somatischen Chromosomen von *Cricetus cricetus* (L.). Chromosoma (Berl.) **18**, 438—448 (1966).

Yerganian, G.: Chromosomes of the Chinese Hamster, *Cricetulus griseus.* I. The normal complement and identification of sex chromosomes. Cytologia **24**, 66—75 (1959).

Chapter 10

Three Different Consequences of X-autosome Translocation

Although they possess the nearly identical DNA value, placental mammals of today display an enormous array of karyotypes, with diploid chromosome numbers ranging from a high of 80 to a low of 17. This indicates that during extensive speciation from a common ancestor, autosomal linkage groups have undergone countless rearrangements. Yet, the original X-chromosome of a common ancestor has apparently been preserved in its entirety by a great majority of placental mammals of today, and in its multiplicated forms by a small number of exceptional species. This conservation of the original X as one unit clearly reveals that the type of X-autosome translocations which split the original X into two separate halves has always been severely deteriorative to speciation, and that these translocations were eliminated as they arose.

Three different consequences of X-autosome translocations discussed in this chapter will show that the dosage compensation mechanism of placental mammals is such that most of the X-autosome translocations are disadvantageous. There is only one type of X-autosome translocation which is compatible with successful speciation and the well being of the species.

a) X-autosome Translocations which Deprive the Female of a Mosaic Status

Searle's X-autosome translocation induced in the mouse *(Mus musculus)* was the type which split the X-chromosome into two halves of nearly equal size (Searle, 1962). The males carrying this translocation were always sterile, but at the time of puberty, there was a burst of transitory meiotic activity by germ cells which made cytological analysis on meiotic figures of the nature of this translocation possible. When various configurations assumed by the XY-autosome quadrivalent were analyzed, it became clear that this was a reciprocal translocation involving the X and an autosome which was considerably smaller than the X. As schematically illustrated in Figure 22, one break occurred near the middle of the X, while the other break split an autosome into a one-third piece and a two-thirds piece (FORD and EVANS, 1964; OHNO and LYON, 1965). Of the two new chromosomes produced by this translocation, one was made of one-half of the X and one-third of an autosome. Consequently, the new chromosome became considerably smaller than the normal X. The other new chromosome, on the other hand, was similar in size to the normal X as it was made of one-half X and two thirds of an autosome (OHNO and LYON, 1965). Detailed genetic analysis of this translocation revealed that a break on the X occurred between the locus for the mutant gene *Tabby* and the locus for the other mutant gene *Blotchy*. Identification of an autosome involved in translocation as the known linkage group was not possible (LYON et al., 1964).

The female balanced for SEARLE's translocation has three chromosomes carrying X-chromosome material: one X in its entirety; and two new chromosomes, each of which carries one-half of the other X. In this circumstance, dosage compensation can be accomplished either by inactivation of the intact X or of the two halves of the other X (Fig. 22). Both cytological and genetical studies revealed that simultaneous inactivation of the two separate halves is beyond the means of the dosage compensation mechanism. Thus, preferential inactivation of the intact X occurred and the female carrying this translocation ceased to be a natural mosaic. When these females were made genetically heterozygous for *Tabby*, they demonstrated either a uniformly *Tabby*, or a uniformly wild-type appearance instead of the customary mosaic phenotype demonstrating vertical stripes of *Tabby* affected areas in the background of wild-type fur. When the intact X carried the mutant gene *Tabby*, the *Tabby* gene was inactivated in all the somatic cells of the body, giving the uniformly wild-type appearance to heterozygous females. Conversely, when the wild-type allele of *Ta* was placed on the intact X, preferential inactivation of the wild-type allele permitted the full expression of *Ta* carried by a translocated X; hence, the uniformly *Tabby* appearance. Preferential inactivation of the allele placed on the normal X was also evidenced when these females were made heterozygous for the other mutant gene *Blotchy*. They were uniformly *Blotchy* when the wild-type allele was placed on the intact X and uniformly wild-type when the intact X carried *Blotchy* (LYON et al., 1964).

Fig. 22. The type of reciprocal X-autosome translocation which deprives the female of her natural mosaic status. The Searle's translocation which occurred in the mouse *(Mus musculus)* is such a translocation. The X-chromosome is painted solid black while the autosome involved is outlined. The inactivated X is shown in the shaded area adjacent to the nuclear membrane. Top: The nature of the translocation. The breaking point on the X occurred in the approximate middle. Middle: The chromosome constitution of a female zygote balanced for this translocation. Bottom: Two theoretical forms of inactivation. Since an X that has split into two pieces cannot be inactivated, the situation shown at the right does not occur. The female then has only one population of somatic cells in which the normal X is inactivated (as shown on the left)

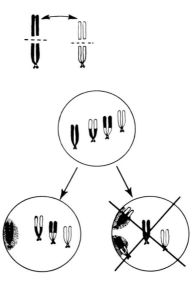

Fig. 22

Since simultaneous inactivation of two separate halves of the X is beyond the means of the dosage compensation mechanism, the complete breakdown of the dosage compensation mechanism is expected to occur in females which became homozygous for this type of X-autosome translocation. When the two X-chromosomes are represented as four separate pieces, it is likely that none of the four would be inactivated.

Obviously, this type of X-autosome translocation is incompatible with successful speciation of placental mammals. This is surely one of the major causes which necessitated the conservation of the original X in its entirety during extensive speciation of placental mammals.

b) X-autosome Translocations which Produce the Variegated-type Position Effect

The variegated-type position effect is in essence the regional suppression of dominant genes which are brought in close proximity to a block of heterochromatin by translocation, inversion or insertion. The recessive alleles then find phenotypic expression in parts of the body of a heterozygote. Such heterozygotes give a mosaic or variegated phenotype with patches of the fully expressed recessive phenotype intermingled with patches of the dominant phenotype.

This phenomenon was first found in the fruit fly *(Drosophila melanogaster)* (MULLER, 1936), and in corn *(Zea mays)* (McCLINTOCK, 1934). In the case of *Drosophila,* when an autosomal segment is inserted next to the heterochromatic region of the X, regional suppression of this autosomal segment occurs irrespective of whether the insertion-bearing X exists alone as in the XY- and XO-males, or is accompanied by the normal X as in the XX- and XXY-females (LEWIS, 1950). This is apparently the reflection of the fact that the concept of "once heterochromatin, always heterochromatin" applies to the heterochromatic region of the *Drosophila* X.

When the type of X-autosome translocation which causes variegated phenotype became known in the mouse *(Mus musculus)* (RUSSELL and BANGHAM, 1959), it became immediately clear that the mechanism of position effect due to X-autosome translocation is basically different from that found in *Drosophila.* In the mouse, the regional suppression of autosomal genes brought adjacent to X-chromosome material depends upon the mosaic nature of the mammalian female. So long as the translocation is carried by the XY-male or XO-female, the translocation-bearing X behaves in an euchromatic manner in all the somatic cells. There would be no suppression of the dominant autosomal genes on a transposed piece. The heterozygotes continue to demonstrate a uniformly dominant phenotype. Only when the translocation is accompanied by a normal X, as in the case of the XX-female and XXY-male, does the translocation-bearing X become heterochromatic in one of the two populations of somatic cells. When the X-chromosome material becomes heterochromatic, an inactivating influence extends to autosomal genes brought adjacent to the X-chromosome material. Parts of the body of a heterozygote populated by the cells with the inactivated translocation-bearing X are now afforded the full expression of autosomal recessive genes which are allelic to the inactivated dominant genes. A variegated or mosaic phenotype results with respect to the autosomal genes.

As much as the manifestation of the variegated-type position effect is dependent upon random inactivation in the female of the normal X

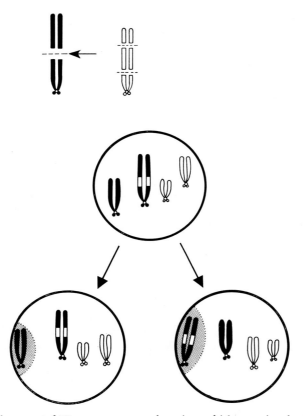

Fig. 23. The type of X-autosome translocation which permits the female to retain her mosaic status, giving rise to the manifestation of variegated-type position effect. Cattanach's translocation which occurred in the mouse (*Mus musculus*) is one such translocation. As shown at the top, the linkage group I autosome has two breaking points, the X-chromosome having one. Translocation results in the insertion of an autosomal piece to the X. The two somatic cell populations of the female balanced for this translocation are shown at the bottom. The nucleus at the right contains the heterochromatinized translocation-bearing X in which inactivation extends to an inserted autosomal piece

or the translocation-bearing X, the type of X-autosome translocation such as SEARLE's, which causes preferential inactivation of the normal X and deprives the female of a mosaic status, cannot possibly result in the manifestation of the position effect. Random inactivation in the female carrying the translocation is possible only if the X-chromosome material as a unit is preserved by the translocation-bearing X. Only an insertion of an autosomal segment to the X or a chromosome break on the X occurring very near to the terminal end does not break the X-chromosome into two separate pieces as a result of an exchange with an autosome. Only these would then result in the manifestation of the variegated-type position effect (Fig. 23).

Of several X-autosome translocations produced by L. B. RUSSELL which manifest the position effect, somatic cells of the two translocations, T (X : 1) 2R1 and T (X : 1) 3R1, were examined by cytological means. An autosome involved in both translocations was the linkage group I, a relatively small autosome. In both, it was found that one of the two new chromosomes which were produced as a result of translocation became larger than the largest pair of autosomes of the normal mouse complement (CHU and RUSSELL, 1965). This great increase in size of the translocation-bearing X is a clear indication that in both translocations a chromosome break occurred very near to the terminal end of the X. Great increase in size of the translocation-bearing X can only be explained by an addition of a linkage group I autosomal material to a nearly entire X. It is our interpretation that RUSSELL's X-autosome translocations result in the variegated-type position effect, because in each a chromosome break occurred at the point very near to either the proximal or the distal end of the X.

In all of RUSSELL's X-autosome translocations, the translocation-bearing male is sterile; this hampers the elucidation of the nature of translocations. In sharp contrast, the variegation-producing X-autosome translocation also involving the linkage group I autosome recovered by CATTANACH (1961) gave semisterile and occasionally even fertile translocation-bearing males. This permitted us to elucidate the exact nature of this translocation by observing meiotic figures of the male. Analysis of configurations presented by the XY-autosome quadrivalent of first meiotic metaphase revealed that one break occurred to the acrocentric X at the point about one-fourth of the total length away from the proximal end. At the same time, two breaks occurred to an autosome of the linkage group I, splitting it into three pieces of about equal size. A middle piece representing about one-third the total length of the linkage group I autosome had been inserted to the X at the point of a break, and there was no reciprocation by the X to an autosome (Fig. 23). As a result of this one-sided exchange, the translocation-bearing X became 20% larger than the normal X, while the size of a linkage group I autosome involved in translocation was reduced by one-third (OHNO and CAT-TANACH, 1962). The genetic study revealed that an inserted autosomal piece contained the gene locus for *chinchilla* and *albino* alleles as well as the gene locus for the *pink-eye* allele.

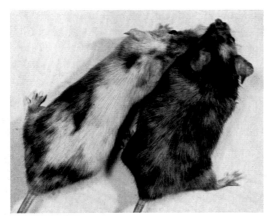

Fig. 24. A photomicrograph of two female mice balanced for Cattanach's translocation, showing the variegation at the *albino* gene locus. The one at the left shows more *albino* patches than the one at the right

When the dominant wild-type allele of *albino* (c^+) was placed on an inserted segment of the translocation-bearing X (X^t), and the recessive gene *albino* (c) was introduced into the gene locus on an intact autosome of the linkage group I, the heterozygous X^tY-male and X^tO-female mice still exhibited a uniformly wild-type appearance. When the X^t carrying the gene c^+ was accompanied by the normal X (X^n), the X^tX^n-females now manifested very clearly the variegated phenotype. These females were flecked with distinct patches of white, revealing regional inactivation of the c^+ allele (Fig. 24). The presence of two populations of somatic cells in these X^tX^n-females was actually proven by direct examination of mitotic prophase figures

(OHNO and CATTANACH, 1962), as well as by examination of metaphase figures labeled with tritiated thymidine (EVANS et al., 1965). In about half of the cells examined, the X^n was in a heterochromatic state and replicated its DNA late. In the other half, it was the larger X^t which was heterochromatic and late labeling. When the X^t was heterochromatic and late labeling, a region which represented an inserted autosomal segment was also heterochromatic and late labeling. Thus, it appeared that the regional suppression in the $X^t X^n$-female of the dominant c^+ allele was due to heterochromatinization of an inserted autosomal segment carrying the c^+ in those melanocytes which had the heterochromatinized X^t. When two X^t's, each carrying the c^+, and two intact linkage group I autosomes, each carrying the c, were brought together into one female, such $X^t X^t$-females exhibited a uniformly wild-type appearance — no patch of white was seen. Random inactivation of one or the other X^t would still leave one functional c^+ in every melanocyte. Obviously, one dose of c^+ produces enough tyrosinase to synthesize the amount of melanin needed for full coloration.

When the $X^t X^n$-females were made heterozygous, not for the c but for another recessive allele *chinchilla* (c^{ch}) at the same gene locus, and the extent of their variegation was compared with that by the other $X^t X^n$-females which were made heterozygous for the recessive allele *pink-eye* (p) at the different gene locus in an inserted segment, the so-called "spreading effect" of inactivation became evident (CATTANACH and ISAACSON, 1965). The mean total area of *pink-eye* patches in the females heterozygous for p was considerably smaller than that of *chinchilla* patches in the females which were variegated with regard to c^{ch}. It appeared that the two gene loci within the same inserted segment were inactivated at a different rate. The most reasonable explanation of the observed spreading effect appears to be that random inactivation of the X^n or the X^t occurs to all the $X^n X^t$-females. As a result, about 50% of the melanocytes of these females are endowed with the inactivated X^t. In the case of the *albino* locus which carries the wild-type allele of two mutant genes c and c^{ch}, the actual inactivation of this gene locus occurs in nearly every melanocyte which contains the heterochromatinized X^t. Thus, the mean total area of c or c^{ch} patches approaches 50% of the total body surface. In the case of the *pink-eye* locus, on the other hand, the complete inactivation of the wild-type allele at this gene locus takes place only in some of the melanocytes with the heterochromatinized X^t. In others, this gene locus escapes the inactivating influence emanated by adjacent heterochromatic X-chromosome material. This accounts for the fact that the mean total area of p patches constitutes considerably less than 50% of the total body surface. It may be that in CATTANACH's X-autosome translocation, one break on an autosome of the linkage group I occurred right next to the *albino* locus so that this gene locus was brought into almost direct contact with the X-chromosome material when insertion took place. On the other hand, the *pink-eye* locus may be situated in the middle of an inserted segment quite a distance away from X-chromosome material on either side. RUSSELL (1963) has found that "spreading effects" also occur in some of her X-autosome translocations, and she has demonstrated that there is a negative correlation between the extent of variegation and the distance between the gene locus and the X-chromosome material. If this is true, the inactivating influence of the heterochromatinized X-chromosome material spreads only to those autosomal genes which are brought into the immediate vicinity of the X-chromosome material.

Alternatively, it is quite probable that autosomal genes with different functions have a different susceptibility to the inactivating influence of heterochromatinized X-chromosome material. The different gene loci concerned with coat color are often mistakenly regarded as having similar functions, but this is far from the truth.

It is becoming clear that the *albino* locus is, in reality, the structural gene locus for an enzyme, tyrosinase. Melanin pigment is a polymer of indole-5,6-quinone which is copolymerized with protein to form melanin granules. The enzyme tyrosinase catalyzes the multistep oxidation of tyrosine to form indole-5,6-quinone. WOLF and COLEMAN (1966) have recently shown that the wild-type (c^+) allele produces two isozymes of tyrosinase, while the mutant *chinchilla* (c^{ch}) allele is a hypomorph, as it produces only one variant isozyme of tyrosinase. The *albino* (c) allele is an amorph which is totally incapable of tyrosinase production. The fact that the c is a recessive gene reveals that the 50% reduction in the tyrosinase level in a heterozygote is still compatible with the full pigmentation of hairs.

The *pink-eye* locus, on the other hand, is not concerned with melanin synthesis. Instead, this gene locus acts on the formation of melanosomes within melanocyte cytoplasma. Each melanosome is made of thin unit fibres which are cross-linked and parallel to each other. When melanin pigment is deposited on this matrix, a melanin granule is formed. The defective allele, *pink-eye* (*p*), causes disorientation of unit fibres. As a result, melanosomes produced under the influence of the *p* are very small in size and extremely irregular in shape. This results in visual dilution of coat color (MOYER, 1963).

As much as the *albino* locus and the *pink-eye* locus have completely different functions, it would not be a great surprise if the two gene loci were to show a different susceptibility to the inactivating influence. At any rate, it becomes quite clear that as long as the X involved is preserved as a whole, X-autosome translocations do not deprive the female of a mosaic status. The inactivating influence emanated by the heterochromatinized X-chromosome material upon autosomal genes which are transposed to the X is limited in scope. Therefore, if semisterility due to the production of genetically unbalanced gametes is avoided, the type of X-autosome translocation which preserves the X as one unit should be quite compatible with the well being of the species and should not hamper further evolution. Indeed, a certain type of X-autosome translocation which occurred during speciation of placental mammals has endured and established the XY_1Y_2/XX scheme of the sex-determining mechanism in certain species.

c) A Type of X-autosome Translocation which Leads to the Establishment of the XY_1Y_2/XX Sex-determining Mechanism

If both chromosome breaks (one on the X and the other on an autosome) occur very near to their respective terminal ends, subsequent exchange would produce a large new chromosome which would be made of nearly one whole of the X and nearly one whole of an autosome. A reciprocal product of such an exchange is so minute in size and made mostly of telomeric and/or centromeric heterochromatin. The loss of it would not cause any serious genetic imbalance (Fig. 25). The male and the female carrying this type of X-autosome translocation would not produce unbalanced gametes — the female is not deprived of a mosaic status. Furthermore, if autosomal genes in the immediate vicinity of the point where the X and an autosome are united are insensitive to the inactivating influence, extensive inactivation of transposed autosomal genes does not occur in the female. Such a translocation shall be maintained by a population, and when the females of the population become homozygous for a translocation, the XY_1Y_2/XX scheme of the sex-determining mechanism emerges. As each new X is actually made of one whole original X and one whole autosome, the XX-female with two doses of the new X is perfectly balanced. In the male, on the other hand, in addition to one new X, the original Y which is now called the Y_1, and a homologue of an autosome which became a part of the new X, are maintained as separate entities. This autosomal homologue is now called the Y_2.

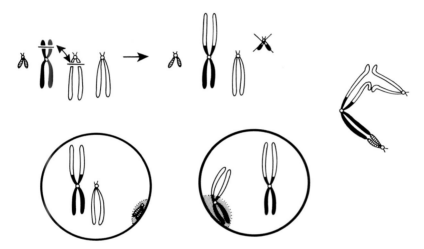

Fig. 25. The type of X-autosome translocation which gives the XY_1Y_2-scheme of the sex-determining mechanism to a species. The original X is painted solid black, the members of an autosomal pair involved in translocation are outlined. The original Y is lined. Top: The nature of translocation. A very terminal translocation attaches a great part of an autosome to the tip of the X. A minute reciprocal product is eliminated. The original Y is now regarded as the Y_1 and an intact autosome as the Y_2, because both of them segregate against the new X during male meiosis (extreme right). Bottom: In the male somatic cell (left), only the Y becomes heterochromatic, while the entire new X remains euchromatic. In the female somatic nucleus (right), only a part of one of the two new X's which represents the original X manifests the heterochromatic condition

During male meiosis, the Y_1 would connect with the original X part of the new X and the Y_2 would pair with an autosomal part of the new X. Unfailing segregation of the Y_1 and the Y_2 against the new X at first meiotic anaphase is assured (Fig. 25). In those species which have the XY_1Y_2/XX mechanism, the diploid chromosome number for the male should be odd and one more than that for the female.

Among mammals, this type of sex-determining mechanism was first convincingly demonstrated in two species of marsupials of Australia: *Potorous tridactylus* ($2n = 13/12$) and *Protemnodon bicolor* ($2n = 11/10$) (SHARMAN and BARBER, 1952; SHARMAN, 1961). In marsupials which belong to the infraclass *Metatheria,* inactivation of one of the two X's of the female (OHNO et al., 1960) occurs in the same manner as in placental mammals which constitute the infraclass *Eutheria,* but the original X of marsupials appears to be considerably smaller than that of placental mammals, suggesting that marsupials and placental mammals did not share an immediate common ancestor (OHNO et al., 1964). HAYMAN and MARTIN (1965) have recently shown that in the female of these two species, only the original X part of one of the two X's shows late DNA replication. An autosomal part of the X seems to escape inactivation completely.

In placental mammals, the XY_1Y_2/XX mechanism has been found in the common shrew of Europe *(Sorex araneus)* of the order *Insectivora* (BOVEY, 1949). In this species, extensive intraspecific autosomal polymorphism of Robertsonian type also operates so that the diploid chromosome number varies from 23 to 31 in individuals (FORD et al., 1957). One species of the gerbils of Israel and North Africa *(Gerbillus gerbillus,* $2n = 43/42$) belonging to the subfamily *Gerbillinae* of the order *Rodentia,* was also found to have the XY_1Y_2/XX constitution (WAHRMAN and ZAHAVI, 1955).

References

BOVEY, R.: Les chromosomes des Chiroptéres et des Insectivores. Rev. suisse Zool. **56**, 341—460 (1949).

CATTANACH, B. M.: A chemically-induced variegated-type position effect in the mouse. Z. Vererb. **92**, 165—182 (1961).

—, and J. H. ISAACSON: Genetic control over the inactivation of autosomal genes attached to the X-chromosome. Z. Vererb. **96**, 313—323 (1965).

CHU, E. H. Y., and L. B. RUSSELL: Pattern of DNA synthesis in X-autosome translocations in the mouse. Genetics **52**, 435 (1965).

EVANS, H. J., C. E. FORD, M. F. LYON, and J. GRAY: DNA replication and genetic expression in female mice with morphologically distinguishable X-chromosomes. Nature **206**, 900—903 (1965).

FORD, C. E., J. L. HAMERTON, and G. B. SHARMAN: Chromosome polymorphism in the common shrew. Nature **180**, 392—393 (1957).

FORD, C. E., and E. P. EVANS: A reciprocal translocation in the mouse between the X chromosome and a short autosome. Cytogenetics 3, 295—305 (1964).

HAYMAN, D. L., and P. G. MARTIN: An autoradiographic study of DNA synthesis in the sex chromosomes of two marsupials with an XX/XY_1Y_2 sex chromosome mechanism. Cytogenetics 4, 209—218 (1965).

LEWIS, E. B.: The phenomenon of position effect. Adv. Genet. 3, 73—115 (1950).

LYON, M. F., A. G. SEARLE, C. E. FORD, and S. OHNO: A mouse translocation suppressing sex-linked variegation. Cytogenetics 3, 306—323 (1964).

McCLINTOCK, B.: The relation of a particular chromosomal element to the development of the nucleoli in Zea mays. Z. Zellforsch. 21, 294—328 (1934).

MOYER, F. H.: Genetic effects on melanosome fine structure and ontogeny in normal and malignant cells. Ann. N. Y. Acad. Sci. 100, 584—606 (1963).

MULLER, H. J.: Variegation in Drosophila and the inert chromosome regions. Proc. Nat. Acad. Sci. 22, 27—33 (1936).

OHNO, S., W. D. KAPLAN, and R. KINOSITA: Basis of nuclear sex difference in somatic cells of the opossum, Didelphys virginiana. Exp. Cell Res. 19, 417—420 (1960).

—, and B. M. CATTANACH: Cytological study of X-autosome translocation in Mus musculus. Cytogenetics 1, 129—140 (1962).

—, C. STENIUS, L. C. CHRISTIAN, W. BEÇAK, and M. L. BEÇAK: Chromosomal uniformity in the avian subclass Carinatae. Chromosoma (Berl.) 15, 280—288 (1964).

—, and M. F. LYON: Cytological study of Searle's X-autosome translocation in Mus musculus. Chromosoma (Berl.) 16, 90—100 (1965).

RUSSELL, L. B., and J. W. BANGHAM: Variegated-type position effects in the mouse. Genetics 44, 532 (1959).

— Mammalian X-chromosome action: inactivation limited in spread and in region of origin. Science 140, 976—978 (1963).

SEARLE, A. G.: Is sex-linked Tabby really recessive in the mouse? Heredity 17, 297 (1962).

SHARMAN, G. B.: The mitotic chromosomes of marsupials and their bearing on taxonomy and phylogeny. Austr. J. Zool. 9, 38—60 (1961).

—, and H. N. BARBER: Multiple sex chromosomes in the marsupial Potorous. Heredity 6, 345—355 (1952).

WAHRMAN, J., and A. ZAHAVI: Cytological contributions to the phylogeny and classification of the rodent genus Gerbillus. Nature (Lond.) 175, 600 —602 (1955).

WOLFE, H. G., and D. L. COLEMAN: Pigmentation. In: Biology of the laboratory mouse, Chapter 21, 2nd edition. Ed. E. L. Green. New York: McGraw-Hill 1966.

Chapter 11

The Consequences of Y-autosome Translocation and the XO as the Normal Female of Certain Mammalian Species

In the previous chapter various consequences of X-autosome translocation were discussed. It was revealed that only one particular type of translocation does not handicap bearers of X-autosome translocation. This type persisted in certain species and produced the XY_1Y_2/XX scheme of the sex-determining mechanism. In this chapter, two unusual sex-determining mechanisms which are produced as a result of Y-autosome translocation shall be discussed.

Because of random inactivation of individual X-chromosomes, the XX-female of mammals actually has only one functioning X in her somatic cells. As the second X of the mammalian female is not really needed, the XO-constitution has indeed become the normal sex chromosome constitution of certain species.

a) The Y-autosome Translocation which Gives an Apparent XO-constitution to the Male

In a great majority of mammals which maintain the original-type X, the Y is a minute element. The very terminal translocation might attach a whole of the Y to the end of a rather large autosome. If the segregation of this Y-carrying autosome from its autosomal homologue as well as from the X is assured during male meiosis, there is no reason why such a translocation can not be maintained by the species (Fig. 26). Since the amount of chromosome material contributed by the Y is very small, the Y-carrying autosome would not look too different from its non-Y-carrying homologue. Thus, until the presence of a trivalent is confirmed in male meiotic figures, the species carrying this type of Y-autosome translocation is apt to be misinterpreted as having the XO/XX sex-determining mechanism.

In the small Indian mongoose *(Herpestes auropunctatus)* belonging to the order *Carnivora*, FREDGA (1965) has recently found that the somatic diploid complement of the male was made of 35 chromosomes, while the diploid number for the female was 36. The single X was readily identified in the male, but the Y was apparently missing. The two X-chromosomes were present in the female. When meiotic figures of the male were examined, however, it was found that the X was always attached at its terminal end to a large autosomal bivalent. The end-to-end association was between the X and the Y which was

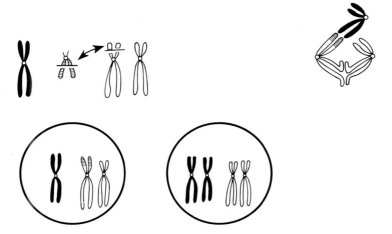

Fig. 26. The type of Y-autosome translocation which gives an apparent XO-constitution to the male. Such a translocation is maintained by the small Indian mongoose *(Herpestes auropunctatus)*. The X is in solid black; the original Y is lined. A pair of autosomes involved in translocation are outlined. Top: A very terminal translocation attaches a great part of the original Y to the very tip of an autosome. A minute reciprocal product of the translocation carrying the Y centromere shall be eliminated. During male meiosis, the X and an intact autosome segregate against an autosome carrying the Y material (extreme right). Bottom: The chromosome constitution of the male (a circle at the left) and the female (a circle at the right). Since the original Y was so minute in size, an autosome carrying the Y material cannot be distinguished from its intact homologue. Thus, the male gives an appearance of having the XO-constitution

attached to the tip of one member of this autosomal bivalent. It must be that at first meiotic anaphase the Y-carrying autosome moves to one division pole, while the X and the non-Y-carrying autosomal homologue moves to the other division pole. This way, the Y-carrying autosome is included in the male-determining gamete, and the X is included in the female-determining gamete. The sex-determining mechanism of the Indian mongoose is, in principle, no different from a great majority of placental mammals.

In a rodent species *Ellobius lutescens* belonging to the subfamily *Microtinae*, MATTHEY (1953) found the diploid chromosome number of 17 in both sexes. The male and female diploid complements were identical with each other. There were eight pairs of autosomes and an odd mediocentric element which was the smallest of the complement. Although it was reported that an odd element existed as a univalent during male meiosis, the most reasonable explanation appears to be that an odd element which constitutes about 5% of the genome is the original-type X of this species. In the male of this species, too, the minute Y is attached to an autosome. This would account for an apparent XO-constitution in the male. The female, on the other hand, may be true XO. In the mouse *(Mus musculus)*, the XO female is normal and fertile (WELSHONS and RUSSELL, 1959). The female *Ellobius lutescens* may have eliminated the unessential second X altogether.

b) Y-autosome Translocation which Produces the $X_1X_2Y/X_1X_1X_2X_2$ Mechanism

In a great majority of placental mammals, the original-type X is accompanied by the minute Y. However, it was noted in the previous chapters that exceptionally large X-chromosomes which are termed the duplicate, triplicate, and quadruplicate-type X's, occur in a small number of rodent species. It appears that duplication and triplication of the original type X were, as a rule, accompanied by a proportional increase in size of the Y. This increase in size of the Y may be due to an addition of X-chromosome material. These large Y's are entirely heterochromatic, therefore X-chromosome material incorporated into the Y should be in a permanently inactivated state.

If exactly the same type of Y-autosome translocation which occurred to the Indian mongoose occurs to these species with the exceptionally large Y, the Y-carrying autosome can be readily distinguished as a new chromosome, for it attained a size much larger than its non-Y-carrying autosomal homologue. Simply because the Y-carrying autosome in this case becomes morphologically distinct, exactly the same mechanism which gave an apparent XO-constitution to the male of the Indian mongoose is now recognized as the $X_1X_2Y/X_1X_1X_2X_2$ mechanism. The true X which was not involved in the translocation is designated as the X_1. Since one Y-carrying autosome is always present in the male, but never in the female, it is called the Y. Its non-Y-carrying homologue occurs unpaired in the male and as a pair in the female. Thus, this autosome which has no bearing

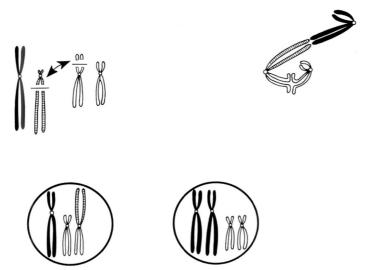

Fig. 27. The same type of Y-autosome translocation which gives the $X_1X_2Y/X_1X_1X_2X_2$-scheme of the sex-determining mechanism to a species. Top: If the species has the duplicate-type X and the proportionately large Y, a very terminal translocation between the Y and an autosome produces a new morphologically distinct chromosome, as an addition of the Y-chromosome material greatly increases the size of an autosome involved in the translocation. This new chromosome is now regarded as the Y, the original duplicate-type X as the X_1, and an intact autosome as the X_2. During male meiosis, the X_1 and the X_2 segregate against the new Y (extreme right). Bottom: The chromosome constitution of the male which is X_1X_2Y (a circle at the left), and the female which is $X_1X_1X_2X_2$ (a circle at the right)

on the act of sex determination is called the X_2 (Fig. 27). At the moment, only one example of the $X_1X_2Y/X_1X_1X_2X_2$ mechanism is available in placental mammals. MATTHEY (1965 a) found this mechanism in one species of African rodent, *Mus minutoides*.

c) The XO as the Normal Female of Certain Species

In the mouse *(Mus musculus)*, the XO-constitution gives rise to a normal and fertile female. One of the two X-chromosomes in female somatic cells of all mammals are rendered inert because of dosage compensation for X-linked genes. It is expected that some mammalian species have achieved an ultimate in dosage compensation by making the XO as the normal female sex chromosome constitution. However, there exist two serious obstacles for the establishment of the XO as the normal female: (1) the Y is essential to maleness, and the presence of a single X is essential to viability; therefore, the male of the species has to maintain the XY-constitution. (Unless something is done, gametes produced by the male carry either an X or a Y.) (2) by becoming the XO, the female sex also becomes the heterogametic sex, producing X-eggs and O-eggs. The double heterogameties are seriously deleterious to the species, for the OY zygote would surely be lethal.

In the creeping vole *(Microtus oregoni)* belonging to the subfamily *Microtinae* of the order *Rodentia,* these two obstacles have been circumvented by developing the mechanism of inducing preferential non-disjunction of the X in primordial germ cells at the right time (OHNO et al., 1963; OHNO et al., 1966). In this species, the male zygote starts its development by having the ordinary XY-constitution ($2n = 18$) and the soma remains the XY, but predirected non-disjunction of the X occurs to primordial germ cells in fetal testis. The XXY- and OY-germ cells are produced. Of these, only the OY's ($2n = 17$) differentiate into definitive spermatogonia. As a result, the male produces two types of sperms, one Y-bearing and the other having no sex chromosome at all. Because the sperm which contains no sex chromosome is the female-determining gamete, the female of this species starts as the XO ($2n = 17$) and the XO-constitution is maintained by the soma. In primordial germ cells of fetal ovaries, however, non-disjunction of the X again takes place. OO-germ cells die off and only the XX-germ cells differentiate into definitive oögonia. This enables the XO-female to produce only one type of egg, each endowed with one X. The double heterogamety is avoided (Fig. 28).

It appears that the female of another rodent species *(Acomys selousi)* from Africa also has the XO-constitution. MATTHEY (1965 b) found the diploid chromosome number of this species to be about 60, with intraindividual variation in the number of autosomes. The male has the ordinary XY-constitution, both in soma and germ lines, but only one X was seen in every diploid complement of the female soma. No information was given as to how this species avoided the hazard of being doubly heterogametic. Since the male germ line maintains the XY-constitution, it must be that the female zygote, in this case, starts its development by having two X's. The elimination of one X during early embryonic development may occur only to the soma, but not to the female germ line.

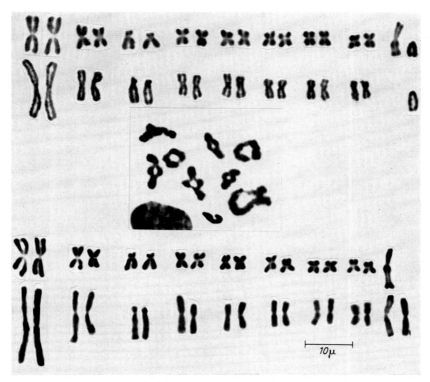

Fig. 28. Photomicrographs illustrating the male and the female of the creeping vole *(Microtus oregoni)* as a gonosomic mosaic. 1) The diploid complement of male somatic cells (the XY-constitution, $2n=18$). The X and the Y are placed at the extreme right. 2) The diploid complement of spermatogonia (the OY-constitution $2n=17$). 3) A diakinesis figure of the male. In addition to eight autosomal bivalents, the presence of the Y univalent (6 O'clock) can be seen. 4) The diploid complement of female somatic cells (the XO-constitution, $2n=17$). 5) The diploid complement of oögonia (the XX-constitution, $2n=18$)

Although not belonging to the infraclass *Eutheria,* six marsupial species of Australia constituting the genera *Isoodon* and *Perameles* appear to have also accomplished an ultimate in dosage compensation

(HAYMAN and MARTIN, 1965). The XO-constitution (2n = 13) was found in somatic cells of both sexes. Mitotic figures from testes, however, had the XY-constitution, and mitotic figures from ovaries, the XX-constitution. In these species, the male zygote obviously starts its development as the XY and the female zygote as the XX. In the case of the male, however, the Y is retained only by germ and somatic elements of the testis. The sole function of the mammalian Y-chromosome is an induction of testicular development; therefore, the presence of the Y is not needed in other tissues. Similarly, in the female, the second X is retained only by germ and somatic elements of the ovary.

If we limit ourselves to placental mammals, it may not be a coincidence that neither of the two species which give the XO-constitution to the female is a possessor of the original-type X. As stated in the previous chapter, the X of *Microtus oregoni* is the triplicate-type, comprising nearly 15% of the genome. MATTHEY's measurement indicated that the X of *Acomys selousi* is also the triplicate-type. Because of its triplicate nature, the single X in female somatic cells of these two species carries, in addition to an euchromatic region which is an equivalent of one whole functioning original-type X, the heterochromatic region which is equivalent to two inactivated original-type X's. In a quantitative sense, the XO-females of these two species have more X-chromosome material than the ordinary females with two original-type X's. This may indicate that the inactivated heterochromatic material is not entirely useless. It is probable that the female of a great majority of placental mammals with the original-type X continues to maintain two X-chromosomes in somatic cells because a subtle contribution from the inactivated second X is not negligible.

With regard to the possible contribution of the heterochromatic X-chromosome material to cell life, the most revealing experiment was carried out by ZAKHAROV and his colleagues (1966). When nine aneuploid clones of fibroblasts derived from the female Chinese hamster *(Cricetulus griseus,* 2n = 22) were studied by the tritiated thymidine labeling technique, it was found on each aneuploid clone that most of the chromosomal material present in excess of the diploid complement was derived from the heterochromatic X chromosome material. It is not likely that the cell would retain totally useless DNA in increased amounts.

We can only conjecture about the role that the genetically inactivated chromosome material plays. However, it should be realized that when we speak of genetic inactivation, we are speaking primarily of the inactivation of structural genes which produce polypeptides of a well defined function. There may be other genetic factors with ill-defined functions which we can not readily comprehend. For example, today we know nothing about the direct products of so-called sex-determining factors. As shall be discussed in detail in Part III, the sex-determining factors appear to be insensitive to the inactivating influence of heterochromatinization.

References

FREDGA, K.: A new sex determining mechanism in a mammal. Chromosomes of Indian mongoose *(Herpestes auropunctatus)*. Hereditas **52**, 411—420 (1965).

HAYMAN, D. L., and P. G. MARTIN: Sex chromosome mosaicism in the marsupial genera *Isoodon* and *Perameles*. Genetics **52**, 1201—1206 (1965).

MATTHEY, R.: La formule chromosomique et le probléme de la détermination sexuelle chez *Ellobius lutescens* Thomas. *Rodentia-Muridae-Microtinae*. Arch. Klaus-Stift. Vererb.-L. **28**, 65—73 (1953).

— Un type nouveau de chromosomes sexuels multiples chez une souris africaine du groupe *Mus (Leggada) minutoides (Mammalia-Rodentia)*. Chromosoma (Berl.) **16**, 351—364 (1965 a).

— Le probléme de la détermination du sexe chez *Acomys selousi* de Winton. Cytogénétique du genre *Acomys (Rodentia-Murinae)*. Rev. suisse Zool. **72**, 119—144 (1965 b).

OHNO, S., J. JAINCHILL, and C. STENIUS: The creeping vole *(Microtus oregoni)* as a gonosomic mosaic. I. The OY/XY-constitution of the male. Cytogenetics **2**, 232—239 (1963).

—, C. STENIUS, and L. CHRISTIAN: The XO as the normal female of the creeping vole *(Microtus oregoni)*. In: Chromosomes today, Vol. I, pp. 182—187. Eds. C. D. Darlington and K. R. Lewis. Edinburgh and London: Oliver and Boyd 1966.

WELSHONS, W. J., and L. B. RUSSELL: The Y-chromosome as the bearer of male determining factors in the mouse. Proc. Nat. Acad. Sci. **45**, 560—566 (1959).

ZAKHAROV, A. F., M. A. EGOLINA, and E. S. KAKPAKOVA: Late-replicating chromosomal segments in aneuploid chinese hamster cell lines as determined by autoradiography. J. Nat. Cancer Inst. **36**, 215—232 (1966).

Chapter 12

Apparent Absence of Dosage Compensation for Z-linked Genes of Avian Species

Although the two are rather similar in absolute size, the avian Z-chromosome makes up nearly 10% of the genome, while the original X of placental mammals comprises only 5%. It would appear that birds have an even greater need than mammals for developing an effective dosage compensation mechanism for their Z-linked genes. It is a great surprise to find that avian species apparently failed in developing an effective means for achieving the dosage compensation.

a) Cytological Behavior of Z-chromosomes in Somatic Cells

In birds, it is the male who is the homogametic sex, having two Z-chromosomes. Therefore, it is expected that if avian species show a sexual dimorphism of somatic interphase nuclei, the sex chromatin body should be seen in the male. Contrary to the expectation, KOSIN and ISHIZAKI (1959) reported that the sex-specific chromocenter was found in nuclei of the heterogametic female sex, while male nuclei were free of a distinct chromocenter. Although their finding has often been disputed by others, it served to focus our attention on the possibility that the behavior pattern of individual Z-chromosomes in somatic cells of birds might be quite different from that of individual X-chromosomes of mammals.

When somatic prophase figures of the male domestic chicken (*Gallus gallus domesticus*) were examined, it was found that neither of the two Z-chromosomes was exhibiting any inclination toward positive heteropyknosis. Both were found in an euchromatic state, assuming a fine thread-like condition (OHNO et al., 1960). Later, SCHMID (1962) studied this problem with the radioautographic technique, labeling chromosomes with tritiated thymidine. Both Z-chromosomes of the male labeled synchronously with the autosomes of comparable size. There was no sign of late replication. The similar thymidine labeling study on the pigeon (*Columba livia domestica*) also revealed the euchromatic nature of both Z-chromosomes in male somatic cells (GALTON and BREDBURY, 1966). It is quite clear that in birds the dosage compensation of Z-linked genes is not accomplished by random inactivation of one or the other Z-chromosomes of the homogametic sex.

When somatic prophase figures of the female chicken were examined, it was felt that one medium-sized chromosome stood out from the rest by virtue of the heterochromatic condition. At that time, the female sex chromosome constitution was thought to be the ZO. Accordingly, this entirely heterochromatic element was interpreted to represent the single Z-chromosome (OHNO et al., 1960). Subsequently, however, the W-chromosome was identified in the female of the domestic chicken (FREDERIC, 1961; SCHMID, 1962), the Australian parakeet (ROTHFELS et al., 1963), the canary and the pigeon (OHNO et al., 1964), and the great horned owl (KRISHAN et al., 1966). Is has now been established that the female sex chromosome constitution of avian species is the ZW and not the ZO. The thymidine labeling study on the female chicken by SCHMID (1962) and that on the female pigeon by GALTON and BREDBURY (1966) revealed that a heterochromatic element in the female is not the Z but the W. The single Z of the female also synthesized its DNA synchronously with the autosomes, but the W was distinctly late labeling. Today, there is no doubt that in somatic cells of both sexes, the avian Z-chromosome invariably behaves in an euchromatic manner along its entire length.

b) Clear Dosage Effect Displayed by Various Avian Z-linked Genes

The cytological observation on the behavior pattern of individual Z-chromosomes indicated that the dosage compensation for avian Z-linked genes is not accomplished by random inactivation of one of the two alleles in the homogametic male sex. If the dosage compensation mechanism evolved in avian species, the mean employed must be very similar to that employed by the fruit fly (Drosophila melanogaster) (see Chapter 7). Genetic evidences indicate that many of the avian Z-linked genes actually show a definite dosage effect. It appears that the dosage compensation mechanism does not exist for many Z-linked genes of the class Aves (COCK, 1964).

A few examples of definite dosage effect exhibited by avian Z-linked genes shall be given here. In the domestic chicken, the barred plumage characteristic of the barred Plymouth Rock breed results from white bands lacking in melanin across feathers that would otherwise be solid black. This barring is caused by a Z-linked gene B.

In the barred Plymouth Rock breed, this gene is maintained in the homozygous *(BB)* state by the male and in the hemizygous *(B —)* state by the female. If there is a dosage compensation for this gene, the male and the female should show an identical plumage. The fact is that males have a lighter plumage than females. Double doses of the *B* cause a wider white band than a single dose of *B*. In fact, the plumage of the hemizygous female *(B —)* with regard to the width of a white band is identical with that exhibited by the heterozygous *(Bb)* male.

The wild-type plumage of the domestic pigeon is slate blue with two black wing bars, a white rump, a black band on its tail, and a green and purple gloss on its neck. This plumage is autosomally determined in absence of an interference from any of a number of Z-linked mutant genes. As stated in Chapter 5, the Z-linked mutant gene *Faded* (B^{of}) dilutes the wild-type plumage to the extreme. The homozygous male with two doses of the B^{of} becomes practically white, while the hemizygous female with a single dose of this mutant gene remains grayish in the same manner as the heterozygous male (LEVI, 1951). The *Barred* of the chicken and the *Faded* of the pigeon, cited above, are but two of many known Z-linked genes of avian species which show a very distinct dosage effect.

There are some Z-linked genes which do not seem to demonstrate a visible dosage effect. In the case of the sex-linked albino gene *(al)* found in the chicken (HUTT, 1949), the Japanese quail (LAUBER, 1964) (Fig. 16), and the turkey (HUTT and MUELLER, 1942), the hemizygous female with a single dose of *al* is as whitish as the homozygous male with two doses. The heterozygous male, on the other hand, maintains a fully colored plumage. An apparent absence of dosage effect, however, does not constitute proof of the existence of the dosage compensation mechanism for the sex-linked albino gene locus, for an amorphic mutant allele is not expected to show the dosage effect under any circumstance.

It is rather unfortunate that today no Z-linked gene with an identifiable direct gene product is known in any of the avian species. If the structural gene for an enzyme, glucose-6-phosphate dehydrogenase, is Z-linked in avian species in absence of the dosage compensation mechanism, it is expected that the enzyme level of the female with one Z is about 50% of that of the male with two Z's. If an amorphic mutant allele exists both hemizygous females and homozygous males are expected to show the zero enzyme level. Hence, an absence of dosage effect for an amorphic mutant allele does not constitute proof that the particular gene locus is compensated for the dosage. A male heterozygous for this amorphic mutant allele, on the other hand, should show the 50% enzyme level seen in the normal female.

Two mutually exclusive interpretations can be given to this apparent absence of a dosage compensation mechanism in avian species. One may conclude that the dosage compensation mechanism for sex-linked genes is a luxury which is desirable to have, but it is definitely not a *sine qua non* of successful speciation. Conversely, one may take the stand that the failure to evolve an effective dosage compensation mechanism for numerous Z-linked genes is one major reason why birds have not escaped the status of feathered reptiles.

c) So-called Splashing Effects Exhibited by Birds Heterozygous or Hemizygous for Z-linked Mutant Genes

In placental mammals, invariably mosaic appearances of the females heterozygous for X-linked coat-color genes have been explained on the basis of a unique dosage compensation mechanism which operates by inactivating one or the other X of the female somatic cells early in embryonic life. As much as the dosage compensation mechanism does not appear to exist for Z-linked genes of avian species, mosaic phenotypes are not expected from male birds heterozygous for Z-linked plumage-color genes. The curious fact is that these heterozygous male and hemizygous female birds quite often demonstrate the so-called splashing effect (HOLLANDER, 1944). It appears that splashing effects are most often observed in the pigeon. The male pigeon homozygous for the Z-linked gene *Almond* (S^t) is practically pure white, while the heterozygous male as well as the hemizygous female demonstrate a yellowish ground color with considerable flecking and splashing of blue, black or brown. If the male carries S^t on one Z-chromosome and the Z-linked gene *brown (b)* on the other Z, he shows brown flecks. It is believed that there are many causes of splashing effects, bipaternity being one of them.

References

Cock, A. G.: Dosage compensation and sex-chromatin in non-mammals. Genet. Res. (Camb.) **5**, 354—365 (1964).

Frederic, J.: Contribution á l'étude du caryotype chez le poulet. Arch. Biol. **72**, 185—209 (1961).

Galton, M., and P. Bredbury: Asynchronous replication of the sex chromosomes of the pigeon *(Columba livia domestica)*. Cytogenetics **5**, 295—306 (1966).

Hollander, W. F.: Mosaic effects in domestic birds. Quart. Rev. Biol. **19**, 285—307 (1944).

Hutt, F. B.: Genetics of the fowl. New York: McGraw-Hill 1949.

—, and C. D. Mueller: Sex-linked albinism in the turkey, *Meleagris gallopavo*. J. Hered. **33**, 69—77 (1942).

Kosin, I. L., and H. Ishizaki: Incidence of sex chromatin in *Gallus domesticus*. Science **130**, 43 (1959).

Krishan, A., G. J. Halden, and R. N. Shoffner: Mitotic chromosomes and the W-sex chromosome of the great horned owl *(Bubo v. virginianus)*. Chromosoma (Berl.) **17**, 258—263 (1966).

Lauber, J. K.: Sex-linked albinism in the Japanese quail. Science **146**, 948—950 (1964).

Levi, W. M.: The pigeon. Columbia, S. C.: The R. L. Bryan Comp. 1951.

Ohno, S., W. D. Kaplan, and R. Kinosita: On the sex chromatin of *Gallus domesticus*. Exp. Cell Res. **19**, 180—183 (1960).

—, C. Stenius, L. C. Christian, W. Beçak, and M. L. Beçak: Chromosomal uniformity in the avian subclass *Carinatae*. Chromosoma (Berl.) **15**, 280—288 (1964).

Rothfels, K., M. Aspden, and M. Mollison: The W-chromosome of the budgerigar, *Melopsittacus undulatus*. Chromosoma (Berl.) **14**, 459—467 (1963).

Schmid, W.: DNA replication patterns of the heterochromosomes in *Gallus domesticus*. Cytogenetics **1**, 344—352 (1962).

Part III

On So-called Sex-determining Factors and the Act of Sex Determination

Chapter 13

Elucidation of So-called Sex-determining Factors

The chromosomal sex-determining mechanism quite obviously depends upon the presence of female-determining factors on one member and of male-determining factors on the other member of the sex chromosome pair. Yet, the exact nature of so-called sex-determining factors remains a riddle. While sex-linked genes are Mendelian genes which behave in a predictable manner, the sex-determining factors are elusive entities which have not permitted close scrutiny.

In this chapter and the next, an attempt shall be made to elucidate the nature of so-called sex-determining factors and their contribution to the actual act of sex determination. No doubt, each sex-determining factor is a particular DNA molecule, but each may exist as multiples on the sex chromosome. They do not appear to be structural genes; rather, they seem to act upon the embryonic indifferent gonad and decide its developmental fate.

Since we are totally ignorant of the nature of individual sex-determining factors of vertebrates, certain inferences have to be made from what is known in the fruit fly *(Drosophila)*, but even in the fruit fly, the knowledge in this regard is very meager indeed. On the other hand, endocrinological information which is essential to our understanding of sex determination is non-existent in the fruit fly, but abound in mammals. The attempt in these two chapters suffers from an indiscriminate assimilation of fragmentary information. Unfortunately, there seems to be no other choice.

a) Relative Potency of the X and the Y in Drosophila and Mammals

Both in *Drosophila melanogaster* and in *Drosophila virilis*, the XO-individual is a male, while the XXY-individual is a female. Thus, it is evident that whether to be a male or a female is not determined by the presence or absence of the Y, but rather by the number of X-chromosomes present. This is due to the fact that the factors which counteract the feminizing influence emanated by the X, are carried not by the Y, but by the autosomes. The diploid set (2A) of autosomes is sufficient to overpower the single X, but is easily subjugated by the combined action of two X's (MORGAN et al., 1925). When the triploid set of autosomes (3A) competes against two X's, the stalemate is reached. Thus, the 3AXXY-individual is an intersex (BRIDGES, 1921). More analytical studies revealed that among three autosomes of *Drosophila*, the third chromosome (second largest autosome) carries most of the male-determining factors (BEDICHEK-PIPKIN, 1959). The Y-chromosome emerges as a "dummy".

In the mouse *(Mus musculus)*, on the contrary, the XO individual is a functioning female (WELSHONS and RUSSELL, 1959); furthermore, in certain other rodent species, the female is normally XO (OHNO et al., 1963; MATTHEY, 1965). In man (JACOBS and STRONG, 1959), and in the mouse *(Mus musculus)* (RUSSELL and CHU, 1961), the XXY is a male, although sterile. On the basis of the above findings, it becomes evident that relative potency of the X and the Y of mammals is quite the opposite of that of the X and the Y of *Drosophila*. Here, the male-determining capacity of the Y-chromosome is without doubt; it is the female-determining capacity of the X which has to be questioned.

In JOST's experiment (1947), removal of the testis in the male fetus of the rabbit automatically resulted in the development of female sexual ducts. On this basis, the view may be taken that in mammals, the female is a neuter sex assumed passively in absence of the Y, and that the 2AXO-constitution is not so much the female constitution, but rather the constitution which is an essential minimum for the individual's development. This, however, appears as an extreme view, for apparent ineffectiveness of the mammalian X as a female-determiner may merely be a reflection of the particular dosage compensation mechanism which inactivates all but one X-chromosome in each somatic cell. Thus, an individual with the 2AXXXXY-constitution still has only one functioning X to counteract the masculinizing influence of the Y. It is no wonder that an individual of this constitution is as much of a phenotypic male as the one with the 2AXXY-constitution.

On the other hand, if the heterochromatinization of extra X-chromosomes completely inactivates female-determining factors on them, the 2AXXY is expected to be a normal male and the 2AXO, a functioning female. The very fact that the 2AXXY-constitution gives rise to a sterile, somewhat feminized male in both the mouse and man, and that the human XO is completely sterile (FORD et al., 1959), reveals that the inactivated sex-chromatin-forming X does influence sexual development. As much as the OO-constitution is lethal due to the fact that the mammalian X carries numerous sex-linked genes which are vital to many metabolic processes but irrelevant to sex determination, we shall never know whether or not mammals can organize a pair of ovaries in absence of the X. Thus, the balanced view appears to be that, in mammals, the Y is very definitely male-determining, but the X does carry some female-determining factors. As of today, we have no information as to the relative potency of the Z and the W of avian species.

b) Sites and Number of Sex-determining Factors

Inasmuch as the X and the Y are chromosomes, one could assume with reasonable certainty that the so-called sex-determining factors are particular DNA molecules which reside on the sex chromosomes. Their mode of action, however, appears to set them apart from ordinary Mendelian genes which we know relatively well.

Each Mendelian gene has a preassigned position on a particular chromosome, and there is a definite dosage effect—a double dose is twice as effective as a single dose. If there is allelic polymorphism, there should be a series of alleles, some hypermorphic and some hypomorphic to the wild-type allele.

In view of our almost total ignorance of the nature of sex-determining genes of mammals and other vertebrates, we shall turn to several revealing experiments done on *Drosophila* to illustrate certain peculiarities of sex-determining factors as genes. As to the nature of the female-determining factors on the X of this insect, most revealing is the partial deletion or duplication induced on the euchromatic region of the X (DOBZHANSKY and SCHULTZ, 1934; PATTERSON et al., 1937). Almost any addition of X-chromosome material pushes the process of sexual differentiation in the female direction and, the longer the added piece, the more pronounced is this effect. Conversely, the balance is tipped in the male direction by deletion of any part of the euchromatic part of the X. These findings suggest that in *Drosophila* the female-determining factors are distributed along the entire euchromatic region of the X, and that the female-determining factors located at one end of the euchromatic region are of the same kind as those located at the other end. It seems that the male-determining factors on the mammalian Y are somewhat similar in nature to the female-determining factors on the *Drosophila* X. It may be that there are only several kinds of male-determining factors on the mammalian Y, but each of these exists in multiples along the entire length of the Y. In man and other mammals, there exists the genuine intraspecific polymorphism of the Y, yet a male with a relatively minute Y does not appear to be less masculine than his fellow with a relatively long Y. The addition of an extra Y, on the other hand, is ineffective in tipping a balance toward the male direction. The 2AXXYY-constitution in man is no better off than the 2AXXY-constitution. An individual with this constitution is still sterile and manifests a definite set of symptoms known as Klinefelter's syndrome (CARR et al., 1961).

Female-determining factors on the *Drosophila* X and male-determining factors on the mammalian Y demonstrate certain apparent peculiarities which set them apart from ordinary Mendelian genes, while some of the male-determining factors of *Drosophila* and one of the female-determining factors on the mammalian X appear to behave more like Mendelian genes.

As described earlier, most of the male-determining factors of *Drosophilia* are located not on the Y, but on the third chromosome (second largest autosome) (BEDICHEK-PIPKIN, 1959). Masculinizing mutant genes have been found on the third chromosome. They may be regarded as hypermorphic alleles of the normal male-determining genes on this chromosome. Both the mutant gene *(ise)* of *Drosophila virilis* (LEBEDEFF, 1939) and the mutant gene *(tra)* of *Drosophila melanogaster* (STURTEVANT, 1945) have no apparent effect on the 2AXX-female in the heterozygous state. In the homozygous state, however, both mutant genes are able to transform a 2AXX-individual into a male, although a sterile one. Thus, some of the male-determining factors of *Drosophila* have definite sites on the third chromosome, and there are hypermorph mutations. In the syndrome of testicular feminization of man, the 2AXY-constitution gives a voluptuously feminine phenotype, although equipped with testes. In the family studies by GAYRAL and his colleagues (1960), this syndrome appears to be due to the hemizygous state of an X-linked mutant gene. It may be that this mutant gene is a hypermorph allelic to one of the normal female-determining genes on the X.

References

BEDICHEK-PIPKIN, S.: Sex balance in *Drosophila melanogaster*. Aneuploidy of short regions of Chromosome 3, using the triploid method. Univ. Texas Publ. **5914**, 69—88 (1959).

BRIDGES, C. B.: Triploid intersexes in *Drosophila melanogaster*. Science **54**, 252—254 (1921).

CARR, D. H., M. L. BARR, and E. R. PLUNKET: A probable XXYY sex-determining mechanism in a mentally defective male with Klinefelter's syndrome. Canad. med. Assoc. J. **84**, 873—877 (1961).

DOBZHANSKY, T., and J. SCHULTZ: The distribution of sex factors in the X-chromosome of *Drosophila melanogaster*. J. Genet. **28**, 349—386 (1934).

FORD, C. E., D. W. JONES, P. E. POLANI, J. C. ALMEIDA, and J. H. BRIGGS: A sex chromosome anomaly in a case of gonadal dysgenesis (Turner's syndrome). Lancet **I**, 711—713 (1959).

GAYRAL, L., M. BARRAND, J. CARRIE et L. CANDEBAT: Pseudo-hermaphrodisme a type de "testicular feminisant": 11 cas. Etude hormonale et etude psychologique. Toulouse Med. **61**, 637—647 (1960).

JACOBS, P. A., and J. A. STRONG: A case of human intersexuality having a possible XXY sex-determining mechanism. Nature **183**, 302 (1959).

JOST, A.: Sur les effects de castration precoce de le embryon male du lapin. C. R. Soc. Biol. **141**, 126—129 (1947).

LEBEDEFF, G. A.: A study of intersexuality in *Drosophila virilis*. Genetics **24**, 553—556 (1939).

MATTHEY, R.: Le probleme de la determination du sexe chez *Acomys selousi* de Winton. Cytogenetique du genre *Acomys (Rodentia-Murinae)*. Rev. suisse Zool. **72**, 119—144 (1965).

MORGAN, T. H., C. B. BRIDGES, and A. H. STURTEVANT: The genetics of *Drosophila melanogaster*. Biblio. Genet. **2**, 1—262 (1925).

OHNO, S., J. JAINCHILL, and C. STENIUS: The creeping vole *(Microtus oregoni)* as a gonosomic mosaic. I. The OY/XY constitution of the male. Cytogenetics **2**, 232—239 (1963).

PATTERSON, J. T., W. STONE, and S. BEDICHEK: Further studies on X-chromosome balance in *Drosophila melanogaster*. Genetics **22**, 407—426 (1937).

RUSSELL, L. B., and E. H. CHU: An XXY-male in the mouse. Proc. Nat. Acad Sci. (Wash.) 47, 571—575 (1961).

STURTEVANT, A. H.: A gene in *Drosophila melanogaster* that transforms females into males. Genetics 30, 297—299 (1945).

WELSHONS, W. J., and L. B. RUSSELL: The Y-chromosome as the bearer of male determining factors in the mouse. Proc. Nat. Acad. Sci. (Wash.) 45, 560—566 (1959).

Chapter 14

Time and Place of Action of Sex-determining Factors in Ontogeny

In multicellular organisms there is a division of labor among the various cell types that constitute an individual. Although endowed with identical genetic constitutions, different somatic cell types in different parts of the body perform different functions. This is because many of the genes are turned on only in certain somatic cell types. The gene loci for component polypeptides of hemoglobin function only in erythropoietic cells. Skin cells and cells of liver parenchyma, no doubt, maintain these genes in their nuclei, yet these genes remain dormant. Similarly, the gene locus for the hormone, insulin, is activated only in Langhans islet cells of the pancreas. The above examples are given in order to demonstrate that every gene with a specialized function has a particular somatic cell type in which to express itself.

For generating energy, the glycolytic cycle and allied pathways are essential to all cells. Accordingly, the enzymes associated with the above mentioned pathways are found in all the somatic cell types of the body. Yet, even in these enzymes, we see evidence of differential activation of gene loci. Lactate dehydrogenase is one such enzyme. In mammals and birds, there are three gene loci for three different component polypeptides of this enzyme. Each enzyme molecule is a tetramer made of four polypeptides. The gene locus for the third polypeptide C is activated only in the testis of male mammals and birds upon sexual maturity and spermiogenesis (GOLDBERG, 1962; BLANCO and ZINKHAM, 1962; BLANCO et al., 1964). Thus, the C4-type lactate dehydrogenase is seen only in the testis containing spermatozoa. In all other tissue of males as well as in all the tissues of the female, we see only five isozymes made of A and B polypeptides in all possible combinations: A4, A3B1, A2B2, A1B3, B4 (MARKERT, 1963). These findings on lactate dehydrogenase not only serve to re-emphasize the fact that each gene has a place of its own to assert itself, but also illustrates another important fact, that each gene has a time of its own to express itself. This latter point is more vividly revealed by the five gene loci for five different component poly-peptides of hemoglobin which is a tetramer. While the gene locus for α-chain remains active throughout the life span of man, the other four gene loci are selectively activated and inactivated in sequence at different developmental stages. When hematopoiesis first begins in the early human embryo, the hemoglobin produced is an embryonic type made of two α-chains and two ε-chains. Subsequently, the fetal hemo-globin which is $\alpha_2\gamma_2$ is produced. The adult hemoglobins, on the other hand, use β- and δ-chains. The two types of hemoglobins are $\alpha_2\beta_2$ and $\alpha_2\delta_2$.

From the above examples, it becomes quite clear that until the time and place of assertion of the sex-determining factors have been delineated, the mere statement that the mammalian Y-chromosome is very strongly male-determining lacks substance.

The question of time and place is particularly relevant in view of the more recent finding on the Y-chromosome of *Drosophila*. In Chapter 13 it was stated that while the mammalian Y is a very strong male determiner, the *Drosophila* Y is a dummy of no consequence. Yet, it has been known that although the XO-constitution of *Droso-phila* gives rise to the male with spermatogenesis, his spermatozoa are incapable of fertilization. Now it appears that the *Drosophila* Y, which remains entirely heterochromatic in all other cell types, unwinds itself in nuclei of primary spermatocytes and becomes euchromatic, asserting its genetic influence. It is assumed that without this asser-tion by the Y in primary spermatocytes, spermatozoa produced are defective (MEYER, 1963). It then follows that the *Drosophila* Y has been misunderstood as a dummy, simply because its time of assertion comes at the very end of gametogenesis. Conversely, it may be said that the mammalian Y emerges as a very strong male determiner only because its time of assertion is at the very beginning of gonadal develop-ment. It is apparent that the mammalian Y behaves as a dummy in most adult somatic cells as well as germ cells. At what time of develop-ment, through which cell type, does the mammalian Y assert its male-determining influence? This question shall be examined step by step.

The sex of an individual is determined very early in embryonic life by the direction of differentiation the indifferent gonad follows; the ovarian differentiation leads to femininity and the testicular differentiation to masculinity. Thus, various cell types which constitute the embryonic gonad become the prime suspects where by sex-determining factors assert themselves. It is obvious that skin, liver, and pancreatic cells, etc., are not concerned with sex determination. Prime suspects are the germ cells themselves and those somatic cells which produce sex hormones.

a) Primordial Germ Cells and their Descendants

In mammals and birds, the site of each future gonad in the early embryo is represented merely as a slight thickening on the surface of mesonephros near the coelomic angle. When this site is stocked with primordial germ cells, the gonadal ridge rapidly grows in size and soon differentiates into either the testis or the ovary. Thus, the sex of an individual is decided fairly early in embryonic life. Even in man, with the rather long gestation period of nine months, this decision is made before the embryo is two months old. It then becomes apparent that the opposing sex-determining factors on the X and the Y must assert themselves at about this time.

As much as the gonadal ridge appears to start its development only after being stocked with primordial germ cells, the question that primordial germ cells might be the cell type through which the sex-determining factors find their expression must be seriously considered.

WITSCHI (1948) was the first to accumulate convincing evidence that germ cells of man and other mammals do not arise *in situ* in the gonad. Instead, they first appear in the upper yolk sac endoderm above the allantoic rudiment when an embryo has but a few pairs of somites. Before the establishment of blood circulation, they begin migration by their own innate locomotion (BLANDAU et al., 1963). During migration toward the gonadal ridge through the hindgut endoderm and the newly formed mesentery, primordial germ cells engage in vigorous mitotic activity and grow in number. Figure 29 illustrates a gonadal ridge of a cattle embryo 15 mm in crown-rump length (about one-month-old) just having been stocked with a full load of primordial germ cells (OHNO and GROPP, 1965).

On the surface, the embryological study on normal gonadal development mentioned above does seem to suggest that it might be the sex chromosome constitution of primordial germ cells which determines

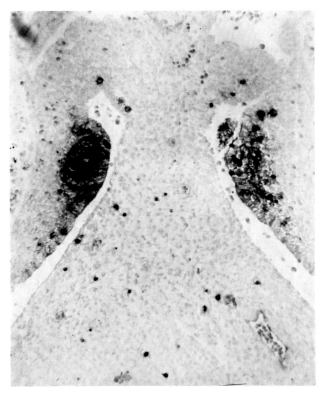

Fig. 29. A transverse paraffin section of a cattle embryo 15 mm in crown-rump length stained for alkaline phosphatase activity. Lense used: 2.5×10. Gonadal ridges are seen on both sides of the dorsal mesentery. Migrating primordial germ cells in dorsal mesentery and coelomic angles are recognized as black spots due to intense alkaline phosphatase activity of their cytoplasma. Many primordial germ cells have already reached the gonadal ridges. As a result, the gonadal ridges are beginning to bulge out

whether the indifferent gonad is to be a testis or an ovary. Indeed, this view has often been expressed, most recently by GOODFELLOW and her colleagues (1965).

However, several revealing findings of somewhat abnormal situations indicate that primordial germ cells themselves play no major role in deciding the future developmental direction of an indifferent gonad.

In the mouse *(Mus musculus)*, there are a series of mutations allelic to the autosomal gene w. Pleiotropic mutant genes W, W^v and W^j, produce an absence of fur pigmentation, anemia, and sterility in a homozygous state. On embryos of W^jW^j-genotype, MINTZ (1957) has shown that sterility of homozygotes is due primarily to the failure of primordial germ cells to propagate during migration. As a result, only very few primordial germ cells reach the gonadal ridges. In the case of WW-homozygotes, deficiency of germ cells is even more severe; the gonads of newborns are totally devoid of germ elements. Yet, a pair of gonads of each individual demonstrates either normal testicular structure or ovarian structure, depending upon the sex chromosome constitution of an individual (COULOMBRE and RUSSELL, 1954).

Vascular anastomosis through fused chorions occurs regularly between dizygotic twin fetuses of cattle (KELLER and TANDLER, 1916; LILLIE, 1917) and the marmoset (WISLOCKI, 1939; BENIRSCHKE et al., 1962). In cattle, this vascular anastomosis occurs while primordial germ cells are still migrating; some of the primordial germ cells appear to wander into vascular circulation (OHNO and GROPP, 1965). As a result, erythrocyte chimerism of dizygotic cattle twins is believed to be accompanied by germ cell chimerism. A number of female (2A-XX) diploid metaphase figures is found in the testes of newborn bulls born twin to the genetic females known as freemartins (OHNO et al., 1962). Histological sections of these newborn testes reveal that 2AXX-germ cells remain within seminiferous tubules as primordial germ cells instead of differentiating into oöcytes and organizing primordial follicles. In cattle, however, these female germ cells transplanted into the testis appear to be unable to differentiate into definitive spermatogonia. Progeny tests done on a number of bulls born twin to freemartins indicate that these bulls transmit only their own blood group genes and not those of their co-twin's (STONE et al., 1960).

In marmosets, on the other hand, transplanted female germ cells appear to be able to engage in spermatogenesis. Not only are female (2AXX) diploid metaphase figures found in the testes of sexually mature males born twin to females, but also a number of first meiotic metaphase figures with the XX-bivalent (BENIRSCHKE and BROWN-

HILL, 1963). Conversely, it may be assumed that 2AXY-primordial germ cells transplanted to the ovary might function as oöcytes. Contrary to the situation found in cattle, in the marmoset and in man (BOOTH et al., 1957), vascular anastomosis between heterosexual twin fetuses does not result in sterilization of the genetic female. The chimeric female born twin to the male is normal and quite fertile.

In normal females of most avian species, only the left gonad functions as an ovary while the right gonad becomes a residual structure essentially made of rudimentary medulla tissue. Atrophy of the ovary caused by tuberculosis, neoplasma, or other pathologic conditions leads to spontaneous masculinization of the female. In the domestic chicken, crowing hens have apparently been the object of curiosity for centuries by the unscientific public. Hence, the proverb says, "A whistling woman and a crowing hen are neither good for gods nor men."

Experimentally, a left ovariectomy in newly hatched chicks leads in many cases to compensatory development of the right residual gonad which becomes a testis (BENOIT, 1923). Small numbers of primordial germ cells which remained in the right residual gonad are now transformed into definitive spermatogonia, and in the fertile seminiferous tubules, spermatogenesis proceeds to the formation of numerous, mature spermatozoa. These spermatogonia, however, maintain the female (2AZW) chromosome constitution (MILLER, 1938).

From the three kinds of revealing findings described above, it may be concluded that although the presence of certain numbers of primordial germ cells might be essential in triggering the growth of the gonadal ridges, primordial germ cells play a very passive role in deciding the developmental direction of embryonic indifferent gonads.

Female primordial germ cells (2AXX in the case of mammals and 2AZW in the case of birds), when placed in a testicular environment, do differentiate into spermatogonia and engage in spermatogenesis. Conversely, this may be true of male primordial germ cells placed in an ovarian environment, although it has not been shown.

The situation found in mice which are made homozygous for an autosomal mutant gene indicates that the differentiation of indifferent gonads either to testes or ovaries can proceed normally in the absence of primordial germ cells. Furthermore, it appears that the post-pubertal ovary can generate a normal estrous cycle without a trace of oöcytes. The mule (2n = 63) is a cross between the male donkey

(Equus asinus, 2n = 62) and the female horse *(Equus caballus,* 2n = 64). There is very little homology between the two parental haploid sets brought together in this hybrid (Trujillo et al., 1962; Benirschke et al., 1962). Consequently, synapsis of homologues is impossible and meiotic cells degenerate at zygotene. In mammals, all the female germ cells enter first meiotic prophase during fetal life; there is no germ cell left in the postnatal ovary of the female mule. Yet, the sexually mature female mule comes into heat regularly and, as shown in Figure 30, a well developed *Corpus luteum* can be seen in her ovary (Ewert, 1899; Benirschke and Sullivan, 1966).

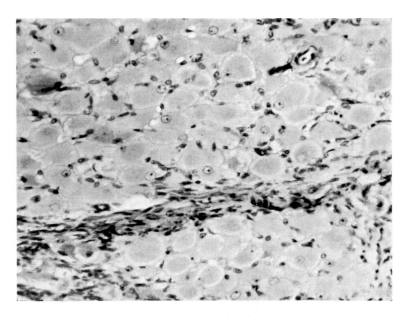

Fig. 30. A part of a *Corpus luteum* from an H and E stained section of the ovary of a sexually mature female mule. The sections were kindly given to the author by Prof. Kurt Benirschke. Lenses used: 25×10

In summary it may be said that germ cells themselves are not concerned with the act of sex determination. In higher vertebrates, germ cells, in a certain sense, have become vassals of somatic elements, particularly those of gonads. Their fate appears to be decided by steroid hormone-producing somatic elements of the gonad.

b) Follicular Cells and Interstitial Cells of the Gonad

The concept of corticomedullary incompatibility with regard to the differentiation of indifferent embryonic gonads has been developed by WITSCHI (1962). In essence, this concept states that the indifferent gonads are hermaphroditic in character. When cortical development suppresses the medulla, the ovary results. While the development of the medullla produces the testis at the expense of the cortex, the stalemate of the corticomedullary competition results in hermaphroditic gonads.

There is much supporting evidence for this concept. In the normally hermaphroditic gonads of the valve snail *(Valvata tricarinata)*, the testicular tissue always occupies the deeper layer of the gonad while the ovarian tissue is situated on the surface (FURROW, 1935). The same topographic arrangement of testicular and ovarian tissues is seen in abnormal hermaphroditic gonads of avian as well as mammalian species. In male toads of the family *Bufonidae,* the testis is invariably capped with ovarian tissue (Bidder's organ), while in female European moles *(Talpa europea),* functioning ovarian tissue is spread over the large testoid structure. In vertebrates, exceptions to this topographic arrangement are found among normally and abnormally hermaphroditic gonads of bony fishes. In cyclostomatas and bony fishes, D'ANCONA (1950) has shown that each gonad develops from a single primordium, and there is no separation into the cortex and medulla at any stage of gonadal development.

The basic premise on which the concept of corticomedullary incompatibility is based is that ovarian follicular cells which are the producer of estrogenic steroid hormones and testicular interstitial cells which are the producer of androgenic steroid hormones are derived from separate progenitors within the indifferent gonad. Follicular cells have been thought to have originated from a strip of peritoneal epithelium which covered the initial gonadal ridge. Hence, the cortical origin of ovarian tissue. Interstitial cells, on the other hand, were thought to have originated from the medial mesonephric blastema situated deep in the center of the gonadal ridge. Hence, the prerequisite for testicular development was thought to be the predominance of medulla over cortex (WITSCHI, 1962). FISCHEL (1929), however, did not believe in the peritoneal epithelium origin of follicular cells, and our more recent observation on fetal ovaries of the cattle also casted the doubt on peritoneal epithelium origin.

Meiotic process of fetal oöcytes should be suspended at diplotene stage and each oöcyte should then be surrounded by a single layer of follicular cells to organize an individual primordial follicle. Thus, successful organization of primordial follicles depends upon a plentiful supply of follicular cells in immediate vicinity of oöcytes. In fetal ovaries, the deeper part of the cortex was endowed with numerous follicular cells, while the area directly beneath the peritoneal epithelium (germinal epithelium) was grossly deficient in follicular cells. As a result, primordial follicles were first formed in the deepest layer of the cortex. Oöcytes in the superficial layer degenerated. This observation certainly suggested that follicular cells originated not from the surface of the gonad but from the bottom (OHNO and SMITH, 1964).

Indeed, the histochemical study of fetal cattle gonads by GROPP and OHNO (1966) revealed the presence of a common blastema for both follicular cells of the female and interstitial cells of the male at the root of the gonadal ridge.

MCKAY and his colleagues (1953) have shown that, in man, the primordial germ cells of early embryos stand out from other cells by virtue of intense cytoplasmic alkaline phosphatase activity. Hence, the histochemical alkaline phosphatase reaction has been utilized to delineate the migration route of primordial germ cells in man, mice, and other mammals. In cattle, it was found that primordial germ cells lose their alkaline phosphatase activity soon after the migration to the gonadal ridges. Subsequently, the progenitors of ovarian follicular cells and testicular interstitial cells make their presence known by the strong enzymatic activity in their cytoplasma. This resulted in the identification of a common somatic blastema in the indifferent gonads of steroid sex hormone-producing cell types.

The entire process of differentiation of a common blastema into ovarian follicular cells, on one hand, and to testicular interstitial cells on the other, is schematically illustrated in Figure 31.

As shown in Fig. 29, the indifferent gonads which have just received their full share of primordial germ cells, are bulging with closely packed, intensely alkaline phosphatase positive cells. Subsequently, as the gonad grows in size, primordial germ cells still positive, are dispersed along the surface of the gonad. At the same time, a mass of cells, showing enzymatic activity as intense as primordial germ cells make their appearance at the radix of the gonad. This is indeed a common blastema to both ovarian follicular cells and testicular interstitial cells.

At this stage no difference in histological structure is detectable between the male and female gonads. The diploid chromosome com-

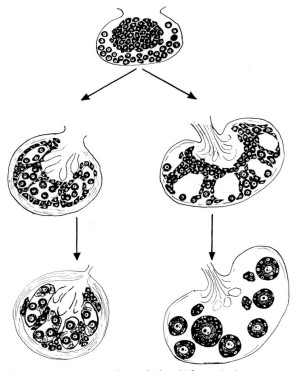

Fig. 31. A schematic representation of the differentiation process of a common somatic blastema of an embryonic indifferent gonad into testicular interstitial cells on one hand, and ovarian follicular cells on the other. Germ cells of all stages are represented as cells with shaded cytoplasma, while testicular interstitial cells, ovarian follicular cells and their common progenitors are represented as cells with black cytoplasma. Top: An indifferent gonad with primordial germ cells in the periphery and the somatic blastema in the center. Middle: The initial stage of sexual differentiation. In the male (left), primordial germ cells actively invade the mass of blastema cells, while blastema cells move slightly toward the periphery, leaving behind the *rete* area in the hilar region. The peripheral area becomes devoid of important elements and organizes the sheath of connective tissue *(Tunica albuginea)*. In the female (right), the blastema sends out strands toward the periphery and meets with primordial germ cells which are then incorporated into strands. Bottom: Now, the testis (left) and the ovary (right) are fully organized. In the testis, the blastema cells are recognized as interstitial cells, while in the ovary, they are recognized as follicular cells

plement of cattle *(Bos taurus,* 2n = 60) is such that the sex chromosome constitution of an early embryo with indifferent gonads can readily be determined by chromosome analysis (OHNO et al., 1962) (Fig. 11 a).

At the next stage, sex differentiation becomes apparent. In the male gonad, primordial germ cells which were distributed along the periphery actively invade the mass of blastema cells and begin to organize seminiferous tubules. Positive blastema cells are left outside the tubules and are now recognized as interstitial cells. At the same time, the mass of blastema cells move out of the hilar region. A phosphatase negative area left behind will become *rete* of the testis. In the female gonad, on the other hand, the majority of primordial germ cells remain directly beneath the surface for some duration and become phosphatase negative. Meanwhile, the blastema begins to send out strands of phosphatase-positive cells toward the surface. These strands represent the follicular cell primordia. When the tip of each strand makes contact with primordial germ cells, the incorporation of germ cells by these strands begins.

The process may be compared with the stuffing of a stocking (follicular cell cords) with oögonia. First, oögonia are engulfed by the peripheral branchings of the cords, then they move downward, deeper and deeper. Finally, only the most remote parts of these cords remain unstuffed. Thus, the cords, hitherto termed follicular cell cords, should now be called ovigerous cords. Within ovigerous cords, oögonia differentiate into oöcytes and enter first meiotic prophase. In the deeper part of the ovigerous cords, follicular cells begin to envelope individual oöcytes in meiosis; thus, primordial follicles are formed. These follicular cells remain intensely alkaline phosphatase positive until near the time of birth. Oöcytes situated directly beneath the surface are not easily reached by follicular cells. Hence, they are unable to organize primordial follicles, but instead degenerate by proceeding too far in the meiotic process. In the medullar part of the ovary, strands of blastema cells which did not receive oöcytes are left as unstuffed cords. It is possible that these unstuffed cords (follicular cell rests) are the elements which enable the female mule ovary to generate a normal estrous cycle and form *Corpus luteum.*

The concept of corticomedullary incompatibility is based on the premise that the follicular cells and the interstitial cells are derived from separate progenitors. In view of the presence of a common blastema to both cell types, this concept appears to be in need of modification. It is not so much the competition between the cortex and medulla, but rather the direction of differentiation followed by the somatic blastema of the gonad which decides the fate of the indifferent gonad.

The male-determining factors on the mammalian Y-chromosome most likely assert themselves in cells of this common blastema and cause them to differentiate into interstitial cells. In the absence of the Y and the invariable presence of the X, follicular cells originate from this blastema. In early embryos, the appearance of this common somatic blastema in the radix of the gonad coincides with the development of another mass of intensely alkaline phosphate positive cells in the area slightly cranial to the gonadal ridge. The latter is found to be a blastema for the adrenal cortex. When serial histological sections of these embryos are examined, it is found that the two blastemas fuse

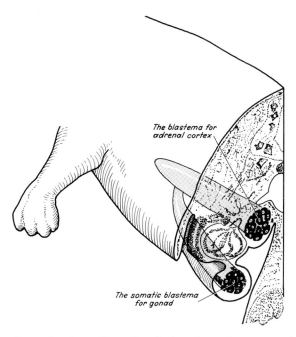

The blastema for
adrenal cortex

The somatic blastema
for gonad

Fig. 32. A schematic three-dimensional drawing of a part of the cattle embryo 27 mm in crown-rump length, demonstrating the topographic relationship between the blastema for gonadal somatic elements and the blastema for the adrenal cortex. The two come in close contact with each other at the junction point, suggesting the common origin

into one at a certain level of the body (Fig. 32). It then appears that there is a common blastema to all three steroid hormone-producing cell types: adrenal cortical cells, ovarian follicular cells, and testicular interstitial cells.

As much as the time and place of action of sex-determining factors have been elucidated, the products of ovarian follicular cells and of testicular interstitial cells should be examined in some detail.

In heterosexual twins in cattle vascular anastomosis between the two invariably modifies the genetic female to a freemartin, which is masculinized and sterile. This masculinization of the genetic female has been attributed to the hormonal influence from the male twin (KELLER and TANDLER, 1916; LILLIE, 1917). However, there had been some doubt as to whether or not fetal testes were capable of producing androgenic steroid hormones. In post-natal life, these hormones are produced in substantial quantity only after puberty. Indeed, the experiments of MACINTYRE (1956, 1960) on transplanted fetal gonads in the rat indicated that substances produced by fetal testes are of a non-steroid type. When a heterosexual pair of fetal gonads on the 16th day of gestation was transplanted together to a subcapsular site in the kidney of a castrated adult rat, an inhibitory action of the testis over the ovary was evident. Parts of the ovary were often transformed into testicular structure. When a distance of more than 8 mm separated the fetal testis and the ovary, no suppressive influence over the ovarian growth was observed. Thus, WITSCHI's notion of "medullarin" (1934) was invoked to explain the observed phenomenon. "Medullarin" was presumed to be a non-steroid hormonal substance produced by the fetal testis.

Even if the existence of "medullarin" is accepted, the extremely short range of its effectiveness rules out the contribution of "medullarin" to freemartinism. In the case of heterosexual cattle twins, testicular influence traverses through vascular channels to affect the ovary of the female twin.

More recently, evidence shows that fetal testes do produce substantial quantities of androgenic steroid hormones. In the marmoset, vascular anastomosis occurs between dizygotic twins with the same regularity as it occurs in cattle, yet the genetic female born twin to the male remains a fertile female (BENIRSCHKE and BROWNHILL, 1962). This is apparently due to the presence of particular enzymes in primate placentae. There is an active enzymatic conversion of androgen to estrogen. Placentae of ungulata obviously lack these enzymes; hence, quantities of androgenic hormones are permitted to enter the body of a female twin fetus (RYAN et al., 1961). Furthermore, the direct assay of fetal human testes revealed the presence of these hormones in appreciable quantities (ACEVEDO et al., 1963).

It is then evident that the interstitial cells of the embryonic and fetal testes already function as the androgenic hormone producer. The critical moment of sex determination is indeed the time of differentiation of a common blastema of the indifferent gonad.

The review of the synthetic pathway of steroid hormones further reveals that whatever the nature of sex-determining factors on the Y and the X-chromosome, their influence manifests itself through manipulation of a metabolic pathway of steroid hormones. The difference between the testis and the ovary with regard to their steroid-hormone-producing capacity is merely quantitative and not qualitative. The three steroid hormone producers of the body are testicular interstitial cells, ovarian follicular cells, and adrenal cortical cells. All use the same metabolic pathway. Gestagenic hormones, androgenic hormones, and estrogenic hormones represent successive steps of one metabolic pathway. The shunt of this main pathway is utilized by the adrenal cortex to produce corticosteroid hormones. Thus, the same metabolic pathway which is used by follicular cells and interstitial cells is also utilized by adrenal cortical cells. Accordingly, certain amounts of androgenic and estrogenic hormones are produced by the adrenal cortex.

It appears that cholesterol molecules should be regarded as a raw material of all the steroid hormones. Cholesterol, in turn, is synthesized from simple acetate molecules. As shown in Figure 33, a direct precursor of steroid hormones is believed to be Pregnenolone (Δ^5-pregnene-3β-ol-20-one) from which Progesterone (Δ^4-pregnene-3,20-dione) is produced. Progesterone, which is essential for the maintenance of pregnancy, is produced in quantity by *Corpus luteum* of the ovary and placenta of pregnant females. At the same time, it is an essential intermediary of androgenic and estrogenic hormones. As such, it is also produced by the testis and the adrenal cortex. Androgenic hormones of the testis are synthesized from progesterone through 17α-hydroxyprogesterone, first Androstenedione (Δ^4-androstene-3, 17-dione) and from it, testosterone (Δ^4-androstene-17β-ol-3-one). Estrogenic hormones of the ovary in turn are synthesized from androstenedione through 19-hydroxyandrostenedione; first, estrone (Δ1,3,5: 10-oestratriene-3-ol-17-one), and from it, estradiol (Δ1,3,5:10-oestratriene-3,17β-diole) and estriol (Δ1,3,5:10-oestratriene-3,16α, 17β-triole). Because of the above relationship between androgenic hormones and estrogenic hormones, the testis normally produces a certain amount of estrogenic hormones as metabolites of androgenic hormones. GOLDZIEHER and ROBERTS (1952) were able to show the presence of 6 µg/kg of estradiol-17β in the human testis, and that the testis of a dog with a feminizing interstitial cell tumor was producing 1000 I. U. of estrogen per gram (LAUFER and SULMAN, 1956). Similarly, the ovary produces certain amounts of androgenic hormones as precursors to estrogenic hormones. The presence of Androstenedione can be detected in the normal ovary. The masculinizing tumor of the ovary can produce a great quantity of androgenic hormones. In the case of the adrenogenital syndrome, the adrenal cortex of both sexes produces a great quantity of androgenic hormones. The primary defect appears to be a deficiency in converting progesterone and/or 17α-hydroxyprogesterone to corticosteroid hormones which are normally the most important products of the adrenal cortex. As a result, ACTH production by the pituitary is not inhibited. The adrenal cortex is subjected to increased stimulation and responds by producing androgenic hormones (BARTTER et al., 1951).

It is clear that the metabolic pathway of steroid hormone production is essentially the same in ovarian follicular cells and testicular interstitial cells and to some extent, in the cells of the adrenal cortex. Yet, the ovary of genetic females is normally geared to produce estrogenic and gestagenic hormones, while the testis of genetic males is normally geared to the production of androgens.

Obviously, the interstitial cells are capable of inhibiting further conversion of androstenedione. The follicular cells, on the other hand, do not permit the accumulation of androstenedione. Instead, androstenedione is converted further to estrone (Fig. 33). The enzymes linked to this metabolic pathway are 3β-hydroxysteroid dehydrogenase, Δ^5-Δ^4-isomerase, 17α-hydroxylase, 11β-hydroxylase, and 21-hydroxylase.

In summary, the male-determining factors on the Y or the Z of vertebrates appear to act first on a common somatic blastema of the embryonic indifferent gonad and direct it to differentiate into interstitial cells, thus forming the testis. When the female-determining factors predominate, a common somatic blastema is directed to differentiate into follicular cells, thus forming an ovary.

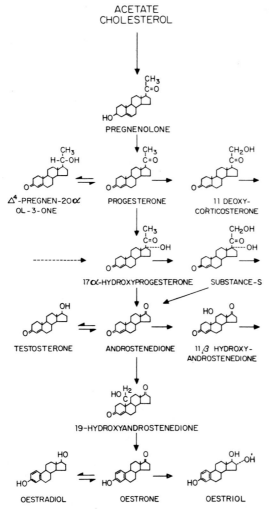

Fig. 33. A scheme of the synthetic pathway of steroid hormones used by the testicular interstitial cells, ovarian follicular cells, and the cells of the adrenal cortex

Whether or not the continuous manipulation by sex-determining factors of a metabolic pathway of steroid hormones is needed to insure the production in quantity of androgenic hormones by interstitial cells and that of estrogenic hormones by follicular cells, is debatable. As stated in Chapter 1, the androgenic hormone treatment during the larval stage can permanently transform the genetic female of fishes and amphibians to the functional male. This ease of sex reversal seems to suggest that the induction of proper interstitial cells automatically ensures the production of androgenic hormones by them. Similarly, in birds, the removal of the left functional ovary of the female causes the right residual gonad to develop as a testis and masculinizes the ZW-female. Apparently, once differentiated into interstitial cells, the ZW-constitution does not interfere with synthesis in quantity of androgenic hormones. On the other hand, it can be argued that what applies to fishes, amphibians, and birds, does not necessarily apply to placental mammals.

A human individual affected by testicular feminization syndrome is the genetic male (XY) endowed with testes, yet the phenotype is feminine equipped with well developed breasts and vagina. Here, the Y was sufficient in directing a common blastema to differentiate into interstitial cells, yet these interstitial cells put out estrogenic rather than androgenic hormones. This can be taken as an evidence that at least in mammals, the induction of interstitial cells *per se* does not automatically ensure the orderly production of androgenic steroid hormones by them.

c) Cells of the Pituitary

The pituitary, a master gland of several other endocrine glands, such as the thyroid and the adrenal, also secretes gonadotropins which influence the activity of the ovary and testis.

It would simplify the matter a great deal if the male pituitary secreted two types of gonadotropins and the female only one of the two types. Gonadotrophins are proteins with some polysaccharides; thus, one type can be the direct product of the male-determining gene on the Y-chromosome, and the other type, the product of the female-determining gene on the X or an autosome. Unfortunately, such does not appear to be the case. Two types of gonadotropins have definitely been identified: follicular stimulating hormone (FSH), and luteinizing hormone (LH). The male as well as the female not only produce both types of hormones, but FSH promotes maturation of Graaffian follicles in the ovary as well as growth of seminiferous tubules in the testis. Similarly, LH which stimulates steroid-genesis by *Corpus luteum* of the ovary is also known as ICSH. In the male, this hormone stimulates steroidgenesis by interstitial cells of the testis (FRIESEN and ASTWOOD, 1965). In the rat, GREEP (1961) has shown that the female functions properly under the influence of the transplanted male pituitary. Clearly, the sex-determining factors do not act through cells of the pituitary.

d) Brief Summary of Time and Place of Action by Sex-determining Factors

It appears that the male-determining factors and the female-determining factors assert their influences through somatic elements of the gonad, and not through germ cells themselves. The first act by the sex-determining factors appears to be in deciding the direction of differentiation to be followed by a common somatic blastema of the indifferent embryonic gonad. The male-determining factors induce it to differentiate into androgen-producing interstitial cells, and the indifferent gonad transforms into the testis. When the female-determining factors predominate, they dictate a common blastema to differentiate into estrogen-producing follicular cells; thus, the ovary is formed. The metabolic pathway of steroid sex hormone synthesis is such that the functional difference between interstitial cells and follicular cells is not qualitative, but quantitative. The former inhibits further conversion of androstenedione, while the latter does not permit the accumulation of androstenedione, but converts it further to estrone.

References

ACEVEDO, H. F., L. R. AXELROD, E. ISHIKAWA, and F. TAKAKI: Studies in fetal metabolism. II. Metabolism of progesterone 4-C^{14} and pregnenolene-7α-H^3 in human fetal testes. J. clin. Endocrinol. Metab. **23**, 885—890 (1963).

BARTTER, F. C., F. ALBRIGHT, A. E. FORBES, A. LEAF, E. DEMPSY, and E. CAROLL: The effects of adrenocorticotropic hormones and cortisone in the adrenogenital syndrome associated with congenital adrenal hyperplasia. An attempt to explain and correct its disordered hormonal pattern. J. clin. Invest. **30**, 237—251 (1951).

BENIRSCHKE, K., and L. E. BROWNHILL: Further observations on marrow chimerism in marmosets. Cytogenetics **1**, 245—257 (1962).

BENIRSCHKE, K., L. E. BROWNHILL, and M. M. BEATH: Somatic chromosomes of the horse, the donkey and their hybrids, the mule and the hinny. J. Reprod. Fertil. **4**, 319—326 (1962).

— — Heterosexual cells in testes of chimeric marmoset monkeys. Cytogenetics **2**, 331—341 (1963).

—, and M. M. SULLIVAN: *Corpora lutea* in proven mules. Fertil. and Steril. **17**, 24—33 (1966).

BENOIT, J.: Transformation experimentale du sexe ovariotomie precoce chez la poule domestique. C. R. Acad. Sci. (Paris) **177**, 1074—1077 (1923).

BLANCO, A., W. H. ZINKHAM, and L. KUPCHYK: Genetic control and ontogeny of lactate dehydrogenase in pigeon testes. J. exp. Zool. **156**, 137—152 (1964).

BLANDAU, R. J., B. J. WHITE, and R. E. RUMERY: Observations on the movements of the living primordial germ cells in the mouse. Fertil. and Steril. **14**, 482—489 (1963).

BOOTH, P. B., G. PLAUT, J. D. JAMES, E. W. IKIN, P. MOORES, R. SANGER, and R. R. RACE: Blood chimerism in a pair of twins. Brit. med. J. I, 1456—1458 (1957).

COULOMBRE, J. L., and E. S. RUSSELL: Analysis of the pleiotropism at the W-locus in the mouse. The effects of W and W^v substitution upon postnatal development of germ cells. J. exp. Zool. **126**, 277—295 (1954).

D'ANCONA, U.: Determination et differenciation du sexe chez le poissons. Arch. Anat. micr. Morph. exp. **39**, 274—294 (1950).

EWERT, J. C.: The Penicuik experiment (Graafian follicle in zebra/horse hybrid). London: Private publication. No publisher 1899.

FISCHEL, A.: Lehrbuch der Entwicklung des Menschen. Berlin: Springer 1929.

FRIESEN, H., and E. B. ASTWOOD: Hormones of the anterior pituitary body. New Engl. J. Med. **272**, 1328—1335 (1965).

FURROW, C. L.: Development of the hermaphrodite genital organs of *Valvata tricarinata*. Z. Zellforsch. mikr. Anat. **22**, 282—304 (1935).

GOLDBERG, E.: Lactic and malic dehydrogenases in human spermatozoa. Science **139**, 602—603 (1962).

GOLDZIEHER, J. W., and I. S. ROBERTS: Identification of estrogen in the human testis. J. clin. Endocrinol. Metabol. **12**, 143—150 (1952).

GOODFELLOW, S. A., S. J. STRONG, and J. S. S. STEWART: Bovine freemartins and true hermaphroditism. Lancet I, 1040—1041 (1965).

GREEP, R. O.: Physiology of the anterior hypophysis in relation to reproduction. Sex and internal secretions (3rd edition). Baltimore: Williams and Wilkins Comp. Volume **1**, 240—301 (1961).

GROPP, A., and S. OHNO: The presence of a common embryonic blastema for ovarian and testicular parenchymal (follicular, interstitial and tubular) cells in cattle, *Bos taurus*. Z. Zellforsch. mikr. Anat. **74**, 505—528 (1966).

KELLER, K., und J. TANDLER: Über das Verhalten der Eihäute bei Zwillingsträchtigkeit des Rindes. Wien. tierärztl. Mschr. **3**, 513—527 (1917).

LAUFER, A., and F. G. SULMAN: Estrogenic Leydig cell tumor with multiple metastases in a dog. The problem of bisexual hormone production by gonadal cells. J. clin. Endocrinol. Metabol. **16**, 1151—1162 (1956).

LILLIE, F. R.: The freemartin; a study of the action of sex hormones in the foetal life of cattle. J. exp. Zool. **23**, 371—422 (1917).

MACINTYRE, M. N.: Effect of the testis on ovarian differentiation in heterosexual embryonic rat gonad transplants. Anat. Rec. **124**, 27—46 (1956).

MACINTYRE, M. N., J. E. HUNTER, and A. H. MORGAN: Spatial limits of activity of fetal gonadal inductors in the rat. Anat. Rec. **138**, 137—148 (1960).

MCKAY, D. G., A. T. HERTIG, E. C. ADAMS, and S. DANZIGER: Histochemical observations on the germ cells of human embryos. Anat. Rec. **117**, 201—220 (1953).

MARKERT, C. L.: Cellular differentiation — an expression of differential gene function. In: Congenital malformations, pp. 163—174 (1964). The International Medical Congress, New York.

MEYER, G. F.: Die Funktionsstrukturen des Y-Chromosoms in den Spermatocytenkernen von *Drosophila hydei*, *D. neohydei*, *D. repleta* und einigen anderen *Drosophila*-Arten. Chromosoma **14**, 207—255 (1963).

MILLER, R. A.: Spermatogenesis in a sex-reversed female and in normal males of the domestic fowl, *Gallus domesticus*. Anat. Rec. **70**, 155—189 (1938).

MINTZ, B.: Embryological development of primordial germ cells in the mouse: Influence of a new mutation W^{jl}. J. Embryol. exp. Morphol. **5**, 396—403 (1957).

OHNO, S., and A. GROPP: Embryological basis for germ cell chimerism in mammals. Cytogenetics **4**, 251—261 (1965).

—, W. D. KAPLAN, and R. KINOSITA: On the end-to-end association of the X and Y-chromosomes of *Mus musculus*. Exp. Cell Res. **18**, 282—290 (1959).

—, H. P. KLINGER, and N. B. ATKIN: Human oögenesis. Cytogenetics **1**, 42—51 (1962).

—, J. TRUJILLO, C. STENIUS, L. C. CHRISTIAN, and R. TEPLITZ: Possible germ cell chimeras among newborn dizygotic twin calves *(Bos taurus)*. Cytogenetics **1**, 258—265 (1962).

—, and J. B. SMITH: Role of fetal follicular cells in meiosis of mammalian oöcytes. Cytogenetics **3**, 324—333 (1964).

RYAN, K. J., K. BENIRSCHKE, and O. W. SMITH: Conversion of androstenedione-4-C_{14} to estrone by the marmoset placenta. Endocrinology **69**, 613—618 (1961).

STONE, W. H., D. T. BERMAN, W. J. TYLER, and M. R. IRWIN: Blood types of the progeny of a pair of cattle twins showing erythrocyte mosaicism. J. Hered. **51**, 136—140 (1960).

TRUJILLO, J. M., C. STENIUS, L. C. CHRISTIAN, and S. OHNO: Chromosomes of the horse, the donkey, and the mule. Chromosoma (Berl.) **13**, 243—248 (1962).

WISLOCKI, G. B.: Obervations on twinning in marmosets. Amer. J. Anat. 64, 445—483 (1939).

WITSCHI, E.: Genes and inductors of sex differentiation in amphibians. Biol. Rev. 9, 460—488 (1934).

— Migration of the germ cells of human embryos from the yolk sac to the primitive gonadal folds. Contr. Embryol. Carneg. Instn. 32, 67—80 (1948).

— Embryology of the ovary. In: The ovary. Baltimore: The Williams and Wilkins Co. 1962.

Author Index

Subject Index

Herstellung: Konrad Triltsch, Graphischer Betrieb, Würzburg